Josiah Conder

The flowers of Japan and the art of floral arrangement

Josiah Conder

The flowers of Japan and the art of floral arrangement

ISBN/EAN: 9783337174880

Printed in Europe, USA, Canada, Australia, Japan

Cover: Foto ©berggeist007 / pixelio.de

More available books at **www.hansebooks.com**

THE

FLOWERS OF JAPAN

AND

THE ART

OF

FLORAL ARRANGEMENT.

THE

FLOWERS OF JAPAN

AND

THE ART

OF

FLORAL ARRANGEMENT

BY

JOSIAH CONDER, F.R.I.B.A.

PROFESSOR OF ARCHITECTURE, AND ARCHITECT TO THE IMPERIAL JAPANESE GOVERNMENT.

WITH ILLUSTRATIONS BY JAPANESE ARTISTS.

TOKIO :

PRINTED AND PUBLISHED BY THE HAKUBUNSHA, GINZA ; ALSO BY KELLY AND WALSH, LIMITED,
YOKOHAMA, SHANGHAI, HONGKONG, AND SINGAPORE.

LONDON : SAMPSON LOW, MARSTON, & COMPANY, LIMITED,
ST. DUNSTAN'S HOUSE, FETTER LANE, E.C.

1892

TO

THE OLD FOLKS AT HOME,

FROM WHOSE COMPANIONSHIP MANY YEARS OF

WORK IN FAR OFF JAPAN HAVE SEPARATED ME,

THIS VOLUME IS AFFECTIONATELY INSCRIBED

BY

THE AUTHOR.

PREFACE.

The interest shown both here and at home in a short paper on the Theory of Japanese Flower Arrangements, read before the Asiatic Society of Japan in 1889, and printed in the Transactions, led me to undertake the preparation of the present work. If, as some reviewers of that essay suggested, a New Art worthy of imitation in the West, was then revealed to them, the present attempt to treat the subject in a fuller and more attractive manner may find its justification.

The results of this endeavour to simplify and recast in more popular shape the principles of the Floral Art of Japan, are doubtless not without numerous imperfections. Some of these are attributable to the pressure of other occupations, and some to the almost insuperable difficulty of rendering clearly in the English language, thoughts and expressions entirely foreign to the Western mind. Other defects may be traced to the lack of local facilities and resources.

No attempt has been made to discuss the different Schools of flower design, each of which lays claim to be the only true exponent of the Art, and to possess secrets unknown to rival teachers. A study of the different theories of these Schools, and of designs illustrated in their published works, shows that the principles of arrangement followed deviate in no important points. Varying terms to describe the same radical lines in compositions, contradictory traditions accounting for practices leading to similar æsthetic results, and a few special instructions for artificially prolonging the vitality of plant and tree cuttings, seem to have been sufficient in many cases, aided by the name of some new patron, to establish an independent School of the Art.

The only modern style of floral design, which, judged by its productions, shows methods of treatment somewhat at variance with the general principles explained in this work, is the *Ikenobo* style, which may be described as a modification of the earliest flower arrangements called *Shin-no-hana* and *Rikkwa*, to which allusion has been made on pages 39 and 40.

In preparing the present work I have received help from several friends to all of whom I beg to tender my best thanks. The botanical works of Professor R. YATABE, of the Imperial University, have been an invaluable aid in classifying the ordinary Japanese plants and trees, and I have also received some kind assistance from that gentleman in identifying some of the rarer specimens. My thanks are also due to Professor BASIL HALL CHAMBERLAIN for translations of Japanese sonnets upon flowers, some of which have been embodied in the first portion of this work. For the principal coloured illustrations, I am beholden to the well-known artist, TSUKIOKA YOSHITOSHI, of the Popular School of painting, the six plates of Flower-viewing being from his brush; also to KAWANABE KIOSUI, daughter of the famous painter KIOSAI, for the illustrations of Japanese interiors. The outline woodcuts are produced from drawings mostly extracted from native works and prepared under my own direction.

The work of printing and publishing has been carried out by the Hakubunsha, Tokio, chiefly under the direction of Mr. MAYEDA RENJIRO, who has besides given much courteous assistance in other ways. I am also under no small obligation to Mr. J. E. BEALE, Manager of the "Japan Mail," for valuable hints as to arrangement and setting up, and to Messrs. F. A. SATOW and W. B. MASON for a careful revision of the proofs.

JOSIAH CONDER.

Tokio, June, 1891.

CONTENTS.

DESCRIPTION OF ILLUSTRATIONS.

COLOURED PLATES.

OUTLINE PLATES.

XII

ILLUSTRATIONS PRINTED WITH THE TEXT.

THE FLOWERS OF JAPAN.

INTRODUCTION.

"Flowers seem intended for the solace of ordinary humanity;
children love them; quiet, tender, cultivated, ordinary people
love them as they grow; luxurious and disorderly people
rejoice in them gathered."—JOHN RUSKIN.

AMONG the many general impressions which exist abroad with regard to Japan, is one, that it is a land abounding in flowers,—that nature has lavished her floral gifts with special favour upon these sunny islands of the Far East. And in a great measure is this popular fancy justifiable, considering the abundant and imposing display produced by flowering trees and shrubs, at certain seasons, in and around the principal cities. In the sense, however, of wild floral profusion, Japanese scenery lacks much that other countries can boast. The traveller from the West, whose rural wanderings nearer home have made him familiar with furze and heather-clad moors, green flower-sprinkled meadows, and hills and forests girdled or carpeted with flowering plants, will miss in Japan some of these charming adornments of natural landscape. There is one short season in the year, when the rice is young, and the honey-scented rape blossom spreads broad stretches of yellow colour over the plains, that recalls to the mind the soft flowering verdure of other lands; but the rice culture, with its endless terraces of mud flats, and the coarse bamboo grass, which in place of softer mead covers every uncultivated hill and glen, deprive the scenery of all but a passing suggestion of the colouring of Western meadows and uplands. The comparative scarcity of groups of wild flowering plants, as a feature in the landscape, is, however, to some extent made up for by the blossoming trees, which at certain seasons show soft masses of colour amid the foliage of the hill sides. The wild camellia, azalia, magnolia, plum, peach, and cherry are the most important of these flowering trees. In particular the wild cherry, which abounds in the Northern island, adds to the wooded landscape an appearance of soft clouds of pale pearly tint, likened to mist upon the mountains.

The flower charm which exists in Japan is not, however, mainly one of pastoral associations, but is closely connected with the national customs and the national art. The artistic character of the Japanese people is most strikingly displayed in their methods of interpreting the simpler of natural beauties. The extravagant taste which demands for its satisfaction the grand, the rare, and the novel, and is unmoved by the modest attractions of more familiar surroundings, is little shared by the masses of the people, who have given their chief sympathies to the humbler aspects of nature. Though the more majestic scenery of the country has served as a constant theme for the painters and poets of the Court and nobility, the popular art has always been an expression of the daily life of the people, and of those simple natural surroundings, customs, and familiar beliefs which contribute to its enjoyment. The recurring months of the year, with the various duties and pleasures which custom has sanctioned, and the quaint fancies with which tradition has clothed them, form the inexhaustible source from which artists of all kinds have looked for inspiration in this country. The common flowers of the seasons have been given a prominent place in the fête-day calendar. Almost every month is known by its special blossoms, and all the important cities have groves and gardens devoted to their public display. Treasured chiefly as heralds of the seasons, and as inseparable from the favourite pursuits and pastimes of out-door life, Japanese flowers are by no means esteemed in proportion to their scarcity or difficulty of production. The isolated merit of rarity, so much sought after in the West, has here little or no attraction. The florists of the country are not deficient in floricultural skill, and produce in certain blossoms forms of considerable artificial exuberance, but the popular taste shows a partiality for the more ordinary and familiar flowers, endeared by custom and association.

Flower viewing excursions, together with such pastimes as *Shell gathering*, *Mushroom picking*, and *Moon viewing*, form the favourite occupations of the holiday seeker throughout the year. By a pretty fancy, the snow-clad landscape is regarded as Winter's floral display, and *Snow viewing* is included as one of the flower festivals of the year. The Chinese calendar, followed until recently by the Japanese, fitted in admirably with the poetical succession of flowers. Spring, the Japanese *Haru*, opened with the New Year, which commenced about February, and was heralded by the appearance of the Plum blossoms.

SPRING FLOWERS.

SPRING FLOWERS.

PLUM BLOSSOMS.

NRICHING the bare landscape with its bloom, and filling the air with its fragrance at a time when the snow of winter has hardly passed away, the blossoming Plum tree has come to be regarded with especial fondness by the Japanese. Combined with the evergreen pine and bamboo, it forms a floral triad, called the *Sho-chiku-bai*, supposed to be expressive of enduring happiness, and is used as a decorative symbol on congratulatory occasions. The Plum blossom is often referred to as the eldest brother of the hundred flowers, being the earliest to bloom in the year. Quick in seizing the peculiar features which distinguish one growth from another, to the extent almost of a tendency to caricature them, the Japanese have been chiefly attracted by the rugged and angular character of the Plum tree, its stiff, straight shoots, and sparse, studded arrangement of buds and blossoms. Thus, a fancy has arisen for the oldest trees, which exhibit these characteristics to perfection. In them is shown the striking contrast of bent and crabbed age with fresh and vigorous youth; and, as if to render more complete this ideal, it is held that the Plum tree is best seen in bud and not in full blossom.

The gardeners of the country, so clever in the training of miniature trees, find in the Plum a favourite object for their skill, imitating in miniature the same character of budding youth grafted on to twisted and contorted age. These tiny Plum trees, trained in a variety of shapes,—bent, curved, and even spiral,—with their vertical or drooping graftings of different coloured blossom-sprays, fresh, fragrant, and long lasting, form one of the most welcome room decorations during the first months of the year.

Poets and artists love to compare this flowering tree with its later rival, the cherry. With the latter, they say, the blossom absorbs all interest, whereas, in the case of the Plum, attention is drawn more to the tree itself: the cherry blossom is the prettier and gayer of the two, but the Plum blossom is more chaste and quiet in appearance, and has,

besides, its sweet odour. The fragrance of the Plum blossom is constantly referred to
in the simple poetry of the country, and the following free translation may be given as
an example of one of such verses :—

> "In Spring time, on a cloudless night,
>   When moon-beams throw their silver pall
>   O'er wooded landscape, shrouding all
> In one soft cloud of misty white,
>   T'were vain, almost, to hope to trace
>   The plum trees in their lovely bloom
> Of argent, t'is their sweet perfume
>   Alone which leads me to their place."

The custom of planting Plum trees in groves and avenues to form pleasure resorts
during blossom time, seems to be of comparatively recent date; and some of the most
famous Plum groves are really orchards, originally planted for the sake of the fruit. It
is said, that in China, from whence Japan borrowed many of her customs and cults, this
tree was first esteemed for its fruit alone, and in later and more æsthetic times it
became honoured for its pure blossom and sweet scent. In the earliest Japanese annals
we read of a single Plum tree being regularly planted in front of the South pavilion
of the Palace at Nara, which was replaced by a cherry tree in later times, when the
latter had in its turn become the favourite of the Court. In connection with this Imperial custom, a pretty story is told explaining the origin of the name *O-shuku-bai*, meaning
Nightingale-dwelling-plum-tree, applied, even to the present day, to a favourite species
with pink double blossom and of delicious odour. Some time in the tenth century the
Imperial Plum tree withered, and, as it was necessary to replace it, search was made for
a specimen worthy of so high an honour. Such a tree was found in the garden of the
daughter of a famous poet, named Kino Tsurayuki, and was demanded by the officials
of the Court. Unable to resist the Imperial command, but full of grief at parting with
her favourite Plum tree, the young poetess secretly attached to its trunk a strip of
paper upon which she wrote the following verse :—

> "Claimed for our Sovereign's use,
>   Blossoms I've loved so long
> Can I in duty fail?
> But for the Nightingale,
>   Seeking her home of song,
> How shall I find excuse?"— BRINKLEY.

This caught the eye of the Emperor, who, reading the lines, enquired from
whose garden the tree was taken, and ordered it to be returned. The season of the

Plum blossom is made musical with the liquid note of the Japanese nightingale, and in the different decorative arts this bird is inseparably associated with the Plum tree. Similar combinations of bird and flower, or even of beast and flower, are numerous, and strictly followed in the many designs of the country; such, for example, are the associations of Bamboo leaves and Sparrows, Pea-fowl and Peonies, and Deer with Maple trees.

In later times Plum trees were planted in large numbers in rural spots near to the ancient Capitals, forming pleasure resorts for the ladies of the Imperial Court. Along the banks of the river Kizu, at a place called Tsuki-ga-se, in the province of Yamato, fine trees of pink and white blossom line the banks for upwards of two miles, diffusing their delicious scent around. These trees are what remains of quite a forest of Plum trees said to have stretched for miles around. The modern Capitals have also their favourite Plum orchards, visited by crowds of sightseers in blossom time, at the end of January. Sugita, a village not far from Yokohama, possesses one of the most famous, having over a thousand trees, many of which are eighty or a hundred years of age, and which supply in the Summer most of the fruit consumed in the Eastern Capital, Tokio. It is popularly known and frequented for its blossoms alone in the early Spring. This orchard boasts six special kinds of tree, distinguished by different fancy names having reference to the character of flower; the principal of which are trees of pink, and others of green blossom,—for the white Plum flower has a faint tinge of emerald. In all, there are said to be sixty different species existing in Japan. The blossom held most in esteem is the single blossom of white or greenish white colour and of small size. All the white kinds are scented, but of the red some have no perfume. There is an early Plum of red double blossom which blooms before the Winter solstice, and is of handsome appearance, but it has little or no scent.

Every visitor to Japan has heard of the *Gwa-rio-bai*, or Recumbent-dragon-plum-trees at Kameido, a famous spot in the North of Tokio. At this place there existed, up to fifty years ago, a rare and curious Plum tree of great age and contorted shape, whose branches had bent, ploughing the soil, forming new roots in fourteen places, and straggling over an extensive area. This tree, from its suggestive shape, received the name of the Recumbent Dragon, and, yearly clad with fresh shoots and white blossoms of fine perfume, attracted large crowds of visitors. From this famous tree fruit was yearly presented to the Shogun. Succumbing at last to extreme age, it has been replaced by a number of less imposing trees, selected on account of their more or less bent and crawling shapes. This present group of trees, inheriting the name and somewhat of the character of Recumbent Dragons, makes a fine show of blossoms in February, and keeps up the popularity of the resort.

Komurai and Kinegawa, near Kameido, also have blossom-groves much frequented.

Another noted spot is Komukai, near Kawasaki, on the Tokaido, not far from the Capital, historically famous as having been often visited by the Shogun, and possessing trees over two hundred years of age.

At Shinjiku, another place on the outskirts of Tokio, is a fine grove of Plums, popularly called the Silver-world *(Gin-sekai)*, a term often applied to the snow-clad landscape, and having special reference in this instance to the silver whiteness of these blossoms.

PLATE I.

VIEWING THE PLUM BLOSSOMS.

CHERRY BLOSSOMS.

THE third month of the old Spring, corresponding with the present April, is the month of the Cherry blossom, the King of flowers in Japan. This flower is remarkable for its softness and exuberance, as contrasted with the severe simplicity of the plum blossom. The latter blooms fresh, vigorous, and leafless, in the bare and often snow-clad landscape; the former, with its florid richness, enhanced in some cases by young reddish leaves, is specially fitted to assert itself amid the greenery of budding Spring. But the splendour of the Cherry's bloom is transitory in comparison with the more lasting qualities of the plum, which retains its beauty for a full month. The Cherry flowers must be viewed during the few short days of their prime, and should these days be stormy, the full glory of the sight is lost. The most enthusiastic partizans of the Cherry blossom assert that it is all the more precious on account of its transient character.

The wild Cherry seems to have existed in Japan from time immemorial, and still abounds in the woods of the Northern island, where the Aino aborigines apply its bark to many purposes. In ancient times, however, the plum tree, of Chinese importation, seems to have absorbed the attention of the Court and people, and it was in later times that the Cherry, the flower of the country, appears to have found its place in their affections. Though early records refer often to the plum, there is no mention of the Cherry earlier than the time of Richiu, an Emperor of the fifth century.

This Monarch was disporting himself with his courtiers in a pleasure boat, on the lake of the Royal park, when some petals from the wild Cherry trees of the adjoining hills fluttered into the wine cup from which he was drinking. This circumstance is said to have drawn His Majesty's notice to the beauty of this neglected blossom, and from this time also arose the custom of wine drinking at the time of Cherry viewing. To this day there is a popular saying, " Without wine who can properly enjoy the sight of the Cherry blossom?" It was reserved for a later Emperor in the eighth century to give to the Cherry that importance as a national flower which it has ever since retained. Whilst on a hunting expedition on Mount Mikasa, in the province of Yamato, the Emperor Shomu, attracted by the beauty of the double Cherry blossoms, composed the following short verse, which he sent, with a branch of the flowers, to his favourite Consort, Komio Kogo :—

> "This gathered cherry branch can scarce convey
> A fancy of the blossom-laden tree,
> Blooming in sunlight; could I shew it Thee,
> Thoughts of its beauty would drive sleep away!"

To satisfy the curiosity of the ladies of His Court, the Emperor afterwards ordered Cherry trees to be planted near the Palace at Nara, and from this time the custom was continued at each succeeding Capital.

Yamato, the province in which were situated several of these ancient Capitals, is the most noted for its Cherry groves, and at a spot called Yoshino a thousand trees line the path and cover the hill side. It was a favourite fancy to compare the appearance of these trees in blossom to mists or snow upon the hills. ("The Cherry blossoms on Mount Yoshino deceive me into thinking they are snow!")

In the thirteenth century the Emperor Kameyama caused a number of Cherry trees from Yoshino to be planted at Arashiyama, a beautiful hilly spot on the banks of the rapid river Oi. Here he built a Summer pavilion, and, in Spring and Autumn, Court after Court visited the lovely spot, rendered further famous in a verse composed by one of the Imperial line ;—"Not second to Yoshino, is Arashiyama, where the white spray of the torrent sprinkles the Cherry blossoms." This spot no longer possesses its Imperial pavilion, but remains a favourite resort for sight-seers from the Western capital, in the months of the Cherry and Maple. Numerous tea houses and booths, on the banks of the rapids, give a fine view of the wooded hills opposite, amidst the Spring greenery of which may be seen the pearly white clouds of the Cherry blossoms. Here the blossoming trees form a part of the distant landscape, as they must have been originally viewed in their natural wildness, when they first attracted the admiration of the earlier Emperors, and before their more gorgeous successors of double-flower became arranged and isolated in artificial orchards and avenues.

In and near to the present Eastern Capital are several spots renowned for their show of Cherry trees in blossom, originally brought from Yoshino, and from the banks of the Sakura river in the province of Hitachi. One of these resorts, at a place called Asukayama, is often spoken of as the New Yoshino.

Koganei, some half day's ride from Tokio, is perhaps the most attractive spot for seeing the double Cherry in full bloom. Here a fine avenue of these flowering trees extends upwards of two and a half miles along the aqueduct which conveys the water of the river Tama to Tokio. It is said that they were first planted immediately after the completion of the aqueduct, by command of the Shogun Yoshimune, in the beginning of the eighteenth century, with the idea that Cherry trees had the virtue of keeping off impurities from water. For this purpose ten thousand trees were brought from Yoshino, and from the banks of the river Sakura; but the number now remaining has dwindled to only a few hundred.

In the old temple grove, now a public park, at Uyeno, there are a number of fine trees of the single early Cherry blossom, called by the Japanese *Higan-zakura*, among which are some magnificent specimens of the Weeping Cherry. This latter species has pendant branches, drooping like the willow, and bears single white flowers, but no fruit; thus being an exception to the general rule, that the trees of single blossom bear fruit whilst those of double blossom are fruitless. The fruit of the Japanese Cherry tree, is, however, at its best insipid and worthless. These trees at Uyeno are all of majestic size, and present a gorgeous sight in April, with their pale pink blossoms seen partly against the blue sky, and partly against the rich foliage of the pines and cedars which surround the golden shrines and cenotaphs of the Shoguns.

The most popular resort in Tokio is the Cherry avenue at Mukojima, extending for more than a mile along the banks of the river Sumida. Here the trees lack the grandeur and natural beauty of those at Uyeno, and have no surrounding foliage to set them off; but they are of double blossom, and bending with their weight of flowers,— looking almost artificial in their luxurious fullness,—present a most imposing sight. This spot is frequented by the gayest holiday makers. Wine drinking is considered essential to a proper enjoyment of the scene, and crowds of pedestrians, bearing their gourds of rice-wine, make such resorts merry and boisterous with their carousals. Other visitors, of a richer class, indulge in the prospect of the blossom-laden banks from roofed pleasure boats, accompanied often by a gay gathering of singing girls.

The month of the Cherry is one of high winds, and the soft petals of the full blown blossoms fall like snow flakes covering the pathways beneath. This simple fact is not without its attraction to the Japanese, who make much of the falling Cherry petal in their poetry and other arts.

> " No man so callous but he heaves a sigh
> When o'er his head the wither'd cherry flowers
> Come flutt'ring down. Who knows ? the Spring's soft show'rs
> May be but tears shed by the sorrowing sky."—CHAMBERLAIN.

PLATE II.

CHERRY BLOSSOMS AT DUSK (YO-ZAKURA). YOSHIWARA, TOKIO.

SUMMER FLOWERS.

SUMMER FLOWERS.

WISTARIA BLOSSOMS.

THE first popular flower of Summer, which, in public places, attracts the pleasure seeker, is the Wistaria, blooming in May, soon after the cherry blossom has fallen. This stalwart flowering creeper is reared upon large trellises, arranged to cover long walks, bridges, or arbours, in pleasure grounds and gardens. A favourite position is one sheltering an open gallery, which overhangs a lake or stream. In the precincts of the popular temple at Kameido, in Tokio, close to the famous plum trees, there are Wistarias of magnificent size, bearing blossoms which hang in rich purple clusters, from two to three feet in length. Wide rustic galleries, in connection with garden *kiosks*, extend over an artificial lake stocked with gigantic gold fish, and the Wistaria trellises form an extended covering overhead. A belief exists that this flower attains great size and beauty if its roots are nourished with the rice-wine of the country, and there is, at Kameido, a tree producing specially fine blossoms, at the base of which visitors are accustomed to empty their wine cups. Other fine specimens exist in various parts of Japan, bearing clusters over three feet in length, among which may be mentioned one at Noda, in the province of Settsu, called the *Shitose*, or tree of a thousand years.

The Wistaria of purple blossom is most common and most esteemed, ranking higher than the white kind. This is an exception to the prevailing custom, which places white before other colours in blossoms of the same species, and especially proscribes purple flowers as associated with mourning, and unfit for felicitous occasions. In various designs the Pheasant is shown in combination with the purple Wistaria.

IRISES.

N June, the popular flower is the Iris or Flag, which is cultivated in large marshy flats near to some river or lake. In many gardens, watered by a stream, a loop or bend in the water-course is spread out into a marshy expanse, planted with Flags, and crossed by fancy plank bridges of zigzag shape. There are four distinct species of Iris, known by different native names, but the kind most seen is the *Acorus calamus*, or Sweet Flag, which the Japanese call *Hana-Shobu*. In the case of displays of Iris flowers, a mass of varied colour is delighted in, the purple, white, and variegated blossoms being grown together, indiscriminately, and with no attempt at pattern or design. The most noted place for shows of this flower is Yatsuhashi, in the province of Mikawa; but the popular resort nearest Tokio is a spot called Horikiri, close to the river Sumida, to which place it is the fashion to made excursions, in pleasure boats, early in June. Here the beds which contain the Flags, in every variety of colour, are surrounded by elevated grassy banks, dotted with summer-houses, from which visitors can look down upon the rich variegated carpet below. Narrow wooden bridges give further picturesqueness to the scene, crowded in the season with a brilliant throng of visitors, whose pretty costumes almost vie in gaiety of colour with the flowers.

The Iris, as a water plant, is associated in art with the Kingfisher, Water Rail, Mandarin Duck, and other water birds.

PLATE III.

WISTARIA BLOSSOMS AT KAMEIDO.

PLATE IV.

VIEWING THE IRISES AT HORIKIRI.

PEONIES AND LOTUS FLOWERS.

SUMMER'S hottest months bring the Peony and Lotus, flowers which, though hardly sufficiently democratic to rank among the most popular, yet play an important part in the art of the country. The Peony is cultivated in long sheltered beds, forming generally the *parterre* to some adjoining chamber, from which its magnificent blossoms can be viewed. In the grounds of the wealthy it is subjected to scrupulous care and nursing, in order to produce flowers of enormous size and fullness, often so large and heavy as to need artificial support. It is regarded as the flower-queen of China, and is essentially the favourite of the upper classes in Japan. The Peony was first, it is said, imported into this country in the eighth century, and was chiefly cultivated in the provinces of Yamato and Yamashiro. Even now, the finest specimens in Tokio are brought from the neighbourhood of the old Capital, Nara. The largest blossoms measure as much as nine inches across. The Peony is sometimes called the Flower-of-prosperity, and another fancy name by which it is know is the Plant-of-twenty-days, given because it is said to preserve its beauty and freshness for that period of time. Of the large Tree-peony there are ninety distinct kinds, not including the small single kind of the same species, of which there are said to exist five hundred varieties. Among colours, the red and white are most valued, purple and yellow specimens, though rare, being less prized. This exuberant flower, with its large curling petals, is a favourite subject for design and decoration. Its companions in art are the Peacock, and the *Shishi*, a kind of conventional Lion, derived from Chinese designs; and in such company it forms the constant decoration of Temple and Palace walls.

The Lotus is closely connected with the Buddhist religion, and is associated, therefore, in the minds of the people, with mortality and spirit-land. The lakes of temple grounds, especially those dedicated to the water goddess Benten, are frequently planted with Lotuses. The fine wide moats of the Tokio Castle abound in these water plants, which impart to them much beauty in the season. Wherever undisturbed pools and channels of muddy water exist, the Lotus is to be found, and even the ditches beside the railway connecting Tokio with its port, are rendered gay in the summer by the Lotus flowers in bloom. As the Peony is said to be the national flower of China, so the Lotus is called the national flower of India, the source and centre of Buddhism. It is therefore considered out of place as a decoration for occasions of festivity and rejoicing, being suggestive of a spiritual life; but it is constantly used for obsequies and other sacred ceremonies. The Lotus serves as a suitable theme for religious contemplation, and is the favourite flower of monastic and temple retreats: the best displays are therefore to be seen in the lakes of the old temple groves of Kioto and other cities. Growing out of the muddiest and most stagnant water, its leaves and flowers are always fresh and clean; although it is particularly sensitive, and quickly withers if brought in contact with any of the foul fertilizers by which other plants are nourished. This purity which the Lotus maintains amid surrounding filth, is mentioned as the reason for associating it with a religious life, and in a well-known book of Buddhist precepts, it is written: —" If thou be born in the poor man's hovel, but hast wisdom, then art thou like the Lotus flower growing out of the mud!"

The white Lotus flower has a powerful and sweet perfume, but the red kind, though more handsome, has little scent. There is a species called the *Gold-thread-lotus*, because its red blossom is marked with yellow lines; and there also exists a very handsome flower of deep crimson colour. The Indian Lotus, which may sometimes be found in Japan, has a large double flower, of red colour, which never closes day or night, but falls off after five or six days. The blossoms of the ordinary kinds close after midday. The leaves of the Lotus are almost as much esteemed as the flowers, their broad curling surfaces of deep green making a fine show in the Lotus ponds, and forming a favourite subject for the painter's brush.

AUTUMN FLOWERS.

AUTUMN FLOWERS.

CHRYSANTHEMUMS.

THE Chrysanthemum is the flower of Autumn, and the triumph of Japanese floricultural skill. Remarkable variety in form and colour of blossom is produced in the specimens cultivated in the gardens of the Court and nobility. The Chrysanthemum flower, in its most exuberant form, loses its disc-like character, and presents a combination of long oval petals, partly extended, and partly curling inwards, exhibiting in contrast the different tints of face and back; whilst, in its most eccentric and artificial shape, it assumes the character of a confused mop of thread-like petals, more curious than beautiful. The florists aim at producing an extraordinary quantity of blossoms upon one stem, reaching often to the number of several hundreds. The varieties of the Japanese *Kiku*, or Chrysanthemum, are numerous, including not only those species classified as such by European botanists, but many kinds of Pyrethrum, Aster, and other generæ.

It seems that the wild Chrysanthemum, of small flower, has always been indigenous to Japan, and held in considerable repute, from very early times, for medicinal purposes; in which connection early records state that large quantities of the yellow flower were yearly sent to the Imperial Court from the Southern provinces. The large cultivated flower, however, is said to have been imported from Corea or China and first planted at Hakata in the province of Chikuzen. At this time five colours were known, described as blue, yellow, red, white, and black, the term black probably referring to a dark purple colour. Originally these plants were reproduced by means of slips and cuttings, but now the seed is employed, which is said to give greater variety of blossom.

The Chrysanthemum is sometimes spoken of as the national flower of Japan, a rank really belonging to the Cherry blossom; and this misconception is probably owing to the former being used as one of the crests of the Imperial House. The flower has always been much honoured by the Court, and as early as the time of the Emperor Heizei, in the ninth century, garden parties were held in the Palace for the purpose of celebrating its blossoming time, just as, at the present day, a yearly Chrysanthemum show takes place in the Imperial grounds. These ancient celebrations seem to have partaken of a truly pastoral character, the Courtiers wearing the plucked blossoms in their hair, drinking wine, and composing verses upon the beauties of the flowers. The modern Chrysanthemum displays in the Palace gardens are more like our own flower-shows in the social conventionality of their arrangements; but the numerous variety, of every imaginable colour and profusion of shape, arranged in long open rustic sheds, forms a brilliant and imposing scene hardly rivalled by any flower-show in the world.

At the last display in the Imperial grounds at Akasaka there were a hundred and sixty varieties of blossom exhibited, each bearing a fancy name of its own. Some of these names are so poetically suggestive of the form or colour of the flowers that the enumeration of a few of them will not be out of place :—

Chi-kin-gi .	Terrestrial Globe—a large globular yellow flower.
Gin-sekai	Silver World—a flower of pure white colour.
Usu-gasumi	Thin Mist—a white flower.
Tsuki-no-tomo . .	Companion of the Moon—a white flower.
Yu-hi-kage	Shadows of the Evening Sun—a flower of dull red colour.
Tama-sudare . .	The Blind of Gems—a flower of orange red colour.
Hatsu-yuki	The First Snow—a pure white blossom.
Hana-gatami	The Basket of Flowers—a rich red flower.
Kagari-bi .	Beacon Light—a red flower.
Asa-hi-no-nami .	Waves in the Morning Sun—a reddish flower.
Ake no-sora	Sky at Dawn—a flower of cherry-blossom colour (pale pink).
Shigarami . .	Garden Fence—a flower the colour of the wistaria blossom (lavender colour).
Asa-ne-gami .	Dishevelled Hair (in morning sleep)—a flower of tangled petals.
Hoshi-dzuki-yo .	Starlight Night—a white flower.
Hoshi-no-hikari .	Star's Brightness—a pale bluish flower.

Kimi-no-megumi . . .		Blessings of Majesty—a pale pink flower.
Yuki-no-ashita		Snowy Morning—a flower of pale pearly pink colour.
Tsuki-no-kasa	.	Moon's Halo—a flower of orange red colour.
Ogon-no-nishiki .	.	Golden Brocade a flower of golden yellow colour.
Shimo-no-ha		Leaves in Frost- a white flower.
Ogon-no-tsuyu		Golden Dew-- a bright yellow flower.
Mangetsu		Full Moon a white flower.
Gek-ka-no-nami . . .		Moonlit Waves—a flower of yellowish white colour.
Haku-rio		White Dragon—a white flower.
Tsuyu-no-shita-zome		Dye of the Dew- a flower of pale yellow colour.

There are said to be in Japan two hundred and sixty-nine colour varieties of the Chrysanthemum, of which sixty-three are yellow, eighty-seven white, thirty-two purple, thirty red, thirty-one pale pink, twelve russet, and fourteen of mixed colours. A fancy prevails that in this flower the same tint is never exactly reproduced, and that in this it resembles the endless variety of the human countenance. Blooming longer than most flowers, the Chrysanthemum has come to be associated with longevity. In the province of Kai, a hill, called the Chrysanthemum Mount, overhangs a river of clear water, into which the petals fall, and a belief exists that long life is assured by drinking the water of this stream. A favourite motive of decoration, which may be seen in numerous conventional designs, is the Chrysanthemum blossom floating in running water. A custom also survives of placing small blossoms or petals in the cup during the wine-drinking which takes place during the festival of the ninth day of the ninth month.

The ordinary varieties of Chrysanthemum are to be seen in great abundance in the street fairs during the Autumn months. Dango-zaka, in Tokio, is the favourite popular resort, and here the flowers are trained into groups of figures and animals, representing historical subjects.

The Chrysanthemum is associated with the Crane, the Royal bird of Japan.

THE SEVEN PLANTS OF AUTUMN.

THE paucity of important flowering trees and plants in Autumn, has, perhaps, led the Japanese to make much of certain simple plants, comparatively insignificant in themselves, but gathering importance and interest in combination. As has already been pointed out, almost every month of the year is associated with a special blossom, and the calendar would not be complete without a reference to these flowers of the late Autumn. These seven plants are: the *Lespedeza*, the Morning Glory, the *Eularia Japonica*, the *Valeriana villosa*, the *Valeriana officinalis*, the *Pueraria Thunbergiana*, and the Carnation. Perhaps the favourite of these is the *Lespedeza*, of which there are several kinds, some having pink, others white, and others yellow flowers. Growing wild on grassy moors, it is associated with Wild Horses, Deer, and the Wild Boar, together with which it is often depicted in different designs. The Deer is specially associated with the Autumn time, and represented also with other Autumn flowers, and with the reddening Maple. The seven Autumn plants are grown together in the *Hyab-kwa-yen*, or Garden-of-a-hundred-flowers, at Mukojima. The temple grounds of the *Hagi-dera*, near Kameido, are famous for their show of *Lespedeza* flowers.

PLATE V.

VIEWING THE SEVEN FLOWERS OF AUTUMN.

MAPLES.

NOTICE of the floral festivals of the year would not be complete without reference to the Maple,—for the reddening leaf of the Maple, like the foliage of many other flowerless trees, is regarded as a flower in Japan. The rich tints of the changing leaves of certain deciduous trees, hardly distinguishable from the colouring of blossoming shrubs such as the azalia, form a favourite object of attraction during the Autumn months. The native term *Momiji*, which is generally translated Maple, is, strictly speaking, a general name applied to many trees which redden in the Fall. Of the Maple itself, there are many varieties, distinguished both by the form of their leaves, and the tone of their colour. No garden is considered complete without its group of Maple trees, placed beside some artificial hill towards the West, to receive additional splendour from the setting sun. Grassy slopes and valleys are planted with these trees, with the object of bringing into one limited prospect the red and golden tints in which the natural scenery of the wooded hills abounds. The grand slopes above the river at Arashi-yama, noted in the Spring time for their show of cherry blossoms, make a fine display of scarlet Maple foliage in the Autumn.

At Ko-no-dai, a famous prominence commanding a view of the whole plain of Tokio, there are some fine Maple trees, noted for their enormous size. A spot called Tatsuta, in the province of Yamato, is renowned for its splendid Maples, which line the banks of the river, and are in full glory about the end of October. At Oji, a suburb of Tokio, the slopes of a natural glen between the hills are thickly planted with fine specimens of these trees, forming a most romantic spot, where, from the galleries of a rustic arbour, may be seen the foliage in all its burning splendour. Shinagawa and Meguro, other well known spots in the vicinity of the Capital, have also good groups of Maple trees which attract many sight-seers. Picnicking and mushroom gathering are pastimes which accompany the viewing of the Maple.

In the poems and pictures of the country the Maple is associated with Deer. A native verse says:—

> "How full of sorrow seems the Autumn! when,
> In solitary rambles slowly straying,
> Amid the russet foliage of the glen,
> I listen to the lonely stag's sad baying."

PLATE VI.

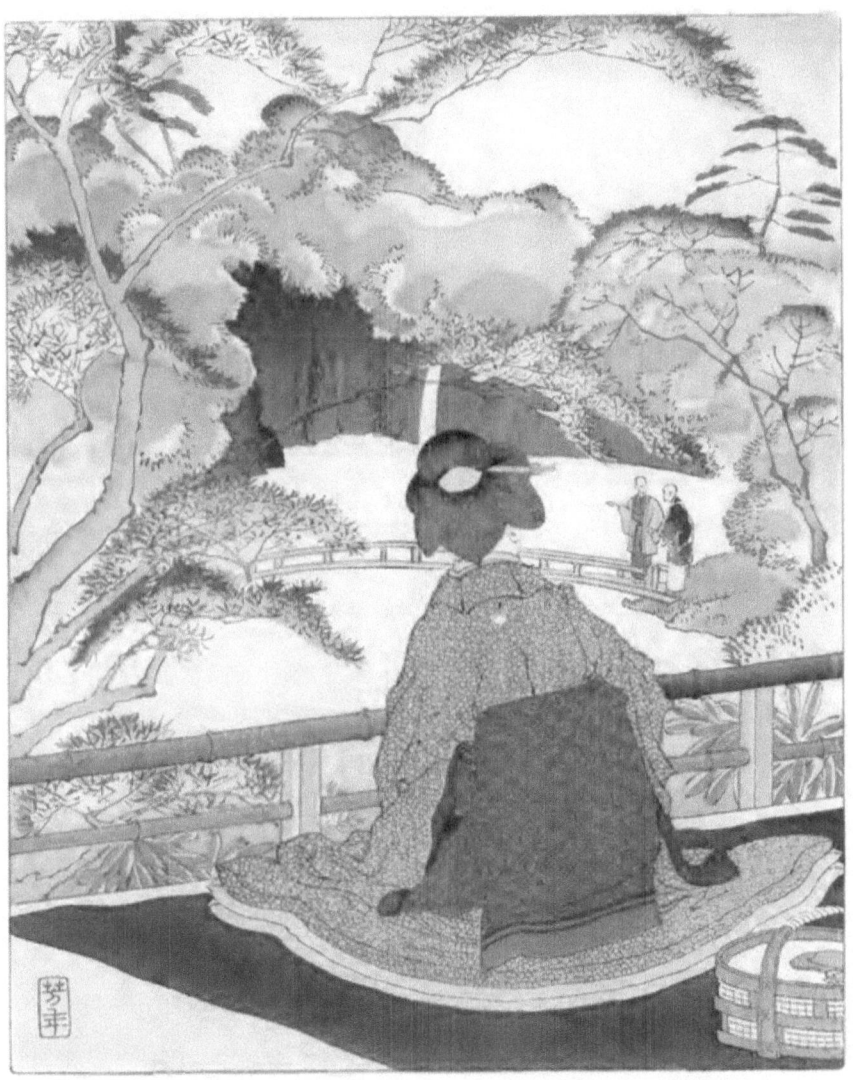

VIEWING THE MAPLES AT OJI.

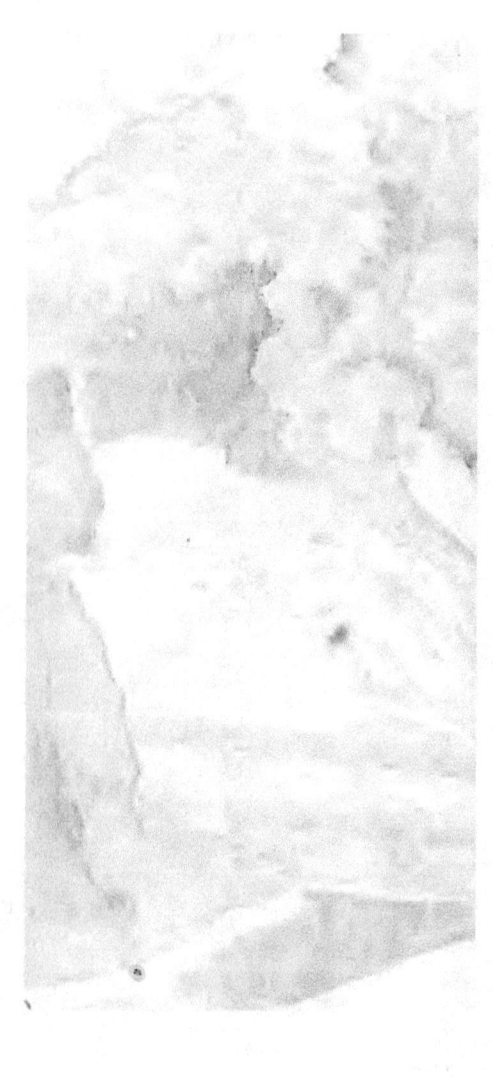

ARRANGEMENT OF FLOWERS.

INTRODUCTION.

THE arrangement of cut flowers in vessels of various kinds has become, with the Japanese, a decorative art of considerable refinement, compared with which the Western methods of floral composition appear but haphazard combinations. The bouquet, the wreath, and the garland, all depending for their beauty upon the close massing of blossoms and greenery in luxurious confusion, bear no resemblance whatever to the more austere and open compositions of the Japanese. The fact that many of the most charming flowers of the country are those of trees, the blossom-clad twigs of which it is impossible to arrange in closed and rounded masses, may in some manner explain the open lineal character given to floral designs; but the same treatment is applied equally to flowering plants and grasses which lend themselves far more easily to the European method of grouping. The peculiarity of treatment noticeable in these flower arrangements is closely connected with the Japanese manner of observing and enjoying floral nature. Whereas the Western amateur devotes his attention mainly to the blossoms, the Japanese lover of flowers bestows his admiration on the whole character of the plant or tree producing them. The rugged nature of the plum trunk, with its straight, stiff shoots, or the graceful sweep of the branches of the weeping cherry, are to him inseparably associated with any beauty which the blossoms themselves possess. The lines of branch and stem, the form and surfaces of leaves, and the distribution of buds and blossoms, all receive their full share of attention. The loveliest buds and blossoms torn from their stems and crushed together in a mass, with ferns or other greenery between them, convey to the Japanese mind no idea of floral art or beauty. The art under consideration is, in fact, based upon a representation, more or less conventional, of floral *growth*; and, principally for this reason, the compositions are made to assume an open character in which the individual forms of branches, stems, leaves, and flowers are all clearly expressed.

The Japanese term *Hana*, translated as *Flower*, has, in the art of floral arrangement, a much wider signification than its English equivalent. Among the so-called *Flowers* of the seasons are included certain evergreens and other flowerless shrubs and trees, some of which hold very high floral rank. The Pine and Bamboo for example, both occupy a very important place in what are called *Flower Arrangements*, and the Maple with its reddening leaves is used as one of the principal *Flowers* of Autumn.

Seasonableness in the choice of material is one of the leading principles which guide the designer. The luxurious taste for choiceness, as associated with the idea of rarity, is diametrically opposed to the rules of the art under consideration. Flowers blooming out of their proper season are, with very few exceptions, rejected for floral compositions, which are intended, in a manner, to be expressive of the particular period of the year. April blossoms, used in any other month, would appear to the flower-artist as incongruous and out of place as would be winter clothing worn in summer-time. It therefore naturally follows that a proper cultivation of the floral art requires a thorough acquaintance with the nature and growth of all trees and plants employed; and even a close observation of the varying characteristics of the same plant during different seasons, in the case of those which are common to several months. The Flag or Iris, for example, which is common alike to certain months of Spring, Summer, and Autumn, has a peculiar bend and vitality in its leaves, and a different length and vigour in its flower stems, at the various periods of its growth; and these distinctions are all kept in view when employing this flower in compositions.

The natural locality of production of trees and plants, whether it be mountain, moor, or lake or river bank, greatly influences the character of the design used. To arrange a water plant in the same manner and with the same feeling and surroundings as a land plant would be considered a great violation of the rules of appropriateness. Not only are blossoming trees and flowering plants treated as perfectly distinct in character, but minor divisions as to locality of production are often observed in both. Among plants a distinction is made between ordinary land plants, forest plants, mountain plants, and water plants; and among trees, land trees, forest trees, and mountain trees are distinguished in some cases.

The imperative necessity for a proper familiarity with the nature of all flowers used in compositions is one reason why it is forbidden to employ those of rare or little known plants, however beautiful they may be. The use of wild flowers, only known to the botanist, as well as rare or foreign flowers with the names of which ordinary folk are not familiar, is prohibited, unless the arranger has previously made himself perfectly acquainted with all the natural characteristics of such flowers. As one exponent of the art has quaintly expressed it, the artist must be thoroughly imbued with a sympathetic feeling for the character, habits, virtues, and weaknesses of the members of the floral kingdom from which he seeks his material, till he possesses almost the same love and tenderness for their qualities as for those of living beings.

Preliminary to a study of Japanese Floral Arrangements some acquaintance with the principal flowers employed is necessary. These flowers are enumerated in the following pages under the heads of the different months to which they belong, and consequently many flowers which are common to several months are repeated. It must be remembered that according to the old calendar the commencement of the first month, which was at the same time the beginning of the Japanese Spring, was about thirty days later than the present first of January. The change of the calendar has therefore rendered it impossible to conform at the present day to all the rules laid down for the selection of flowers for special occasions. Such of the old fête-days as are now observed, being pushed back one month or more in time, the flowers originally fixed as appropriate for their celebration are often unavailable, or recourse has to be made to premature or forced specimens. The following classification is according to the old calendar, which ruled when the whole theory of the art in question was established.

6

FLOWERS ACCORDING TO THEIR MONTHS

(OLD CALENDAR).

1st MONTH (PRESENT FEBRUARY).

Japanese Name.	Botanical Name.	English popular Name.	Japanese Name.	Botanical Name.	English popular Name.
*Fuku-ju so	Adonis amurensis		*Rengio	Forsythia suspensa	
§Suisen	Narcissus Tazetta	Narcissus	Tsubaki	Camellia japonica	Camellia
*Uguisu-so	Lithospermum Zollingeri		*Murozaki-Momo	Jasminum Sieboldianum	Hot-house Peach
*Hakubai	Prunus mume	White Plum	*Obai	Jasminum Sieboldianum	
§Yanagi	Salix japonica	Willow	*Kinsenkwa	Calendula officinalis	
§Kan-giku	Pyrethrum sinense	Winter Chrysanthemum	*Choshun	Rosa indica	
			*Mansaku	Hamamelis japonica	
*Yabu-koji	Ardisia japonica		§Robai	Chimonanthus fragrans	Chinese Plum

---

2nd MONTH (PRESENT MARCH).

Japanese Name.	Botanical Name.	English popular Name.	Japanese Name.	Botanical Name.	English popular Name.
‡Hakubai	Prunus mume flore alba	White Plum	*Tennan-sho	Arisæma japonicum	
‡Hi to		Red Peach	†Oka kotono	Nuphar japonicum	
‡Obai	Jasminum Sieboldianum		*Rengio	Forsythia suspensa	
*Ko bai	Prunus mume flore rosa	Red Plum	*Anzu	Prunus armeniaca	Apricot
*Higan zakura	Prunus subhirtella		‡Hitoye-zakura	Prunus pseudo cerasus	Single Cherry
*Usu to	Prunus persica	Pale Peach	†Yamabuki	Kerria japonica	
*Niwatoko	Sambucus racemosa		*Hitsuji-so	Nymphæa tetragona	
*Kuma-so	Dicentra spectabilis		*Niwa-ume	Prunus japonica	Garden Plum
‡Azuma-giku	Erigeron Thunbergii		*Wase-zakura	Prunus pseudo cerasus	Early Cherry
‡Kinsenkwa	Calendula officinalis		*Niwa-zakura	Prunus pseudo cerasus	Garden Cherry
†Haru giku	Chrysanthemum coronarium	Spring Chrysanthemum	‡Enishida	Cytisus scoparius	
*Hotei chiku	Bambusa sterilis		‡Moku-renge	Magnolia	Magnolia

Japanese Name.	Botanical Name.	English Popular Name.	Japanese Name.	Botanical Name.	English Popular Name.
Suwo	Caesalpinia Sappan		Nashi	Pyrus ussuriensis	Pear
Yobai-kwa	Myrica rubra		Sumomo	Pyrus triflora	
Shakunage	Rhododendron Metternichii	Azalea	Ringo	Pyrus malus	Apple
Tsutsuji	Rhododendron indicum	Azalea	Asebo	Andromeda japonica	
Kaido	Pyrus spectabilis		Sendai-hagi	Thermopsis fabacea	
Obai	Jasminum Sieboldianum		Korai-giku		Corean Chrysanthemum
Ranan	Aspidistra lurida		Boke	Pyrus japonica	
Bijin-so	Papaver rhœas		Mansaku	Hamamelis japonica	
Uguisu-so	Lithospermum Zollingeri		Choshun	Rosa indica	
Yuki-wari-so	Anemone hepatica		Wasure-gusa	Hemerocallis flava	
Kaza-omodaka	Alisma plantago		Ita-dori	Polygonum cuspidatum	
Shun-ran	Cymbidium virens		Sumire	Viola Patrinii	
Kobushi	Magnolia kobus	Magnolia	Uikio	Fœniculum vulgare	
Tsubaki	Camellia japonica	Camellia			

3rd MONTH (PRESENT APRIL).

Japanese Name.	Botanical Name.	English Popular Name.	Japanese Name.	Botanical Name.	English Popular Name.
Haku-to	Prunus persica flore alba	White Peach	Shakunage	Rhododendron Metternichii	Azalea
Usu-to		Pink Peach	Yamamomo	Myrica rubra	
Nojiro-momo		Peach	Asebo	Andromeda japonica	
Hi-to	Prunus persica	Red Peach	Sendai-hagi	Thermopsis fabacea	
Nora-momo		Peach	Fuji-kazura		
Hosumomo		Peach	Shakuyaku	Pæonia albiflora	Peony
Ko-to	Prunus persica flore rosa	Red Peach	Kakitsubata	Iris lævigata	Iris
Gempei-momo		Red and White Peach	Azami	Argemone mexicana	
Rito	Prunus triflora		Ayame	Iris sibirica	Iris
Nashi	Pyrus ussuriensis	Pear	Karamatsu-so	Thalictrum aquilegifolium	
Ringo	Pyrus malus	Apple	Tessen	Clematis florida	Clematis
Yamabuki	Kerria japonica		Wasure-gusa	Hemerocallis flava	
Rengio	Forsythia suspensa		Togiri	Clerodendron squamatum	
Kobushi	Magnolia kobus	Magnolia	Keshi	Papaver somniferum	
Anzu	Prunus armeniaca	Apricot	Kodemari	Spiræa cantoniensis	
Haru-giku	Chrysanthemum coronarium	Spring Chrysanthemum	Ippatsu	Iris tectorum	Iris
Jinchobe	Daphne odora	Daphne	Shaga	Iris japonica	Iris
Ebine-so	Calanthe discolor	Kind of Orchid	Botan	Pæonia moutan	Tree Peony
Kume-zakura	Spiræa Thunbergii		Awamori-so	Astilbe japonica	
Niwa-zakura	Prunus pseudo-cerasus	Garden Cherry	Kazaguruma	Clematis patens	Clematis
Suwo	Caesalpina happan		Shiran	Bletia hyacinthina	
Wase-cukura	Prunus pseudo cerasus	Early Cherry	Choshun	Rosa indica	
Korai-giku		Corean Chrysanthemum	Suzuran	Convallaria majalis	
Mokuren	Magnolia	Magnolia	Tsurigane-so	Campanula punctata	Bluebell
			Konniaku	Conophallus konjak	
			Enishida	Cytisus scoparius	

Japanese Name.	Botanical Name.	English popular Name.	Japanese Name.	Botanical Name.	English popular Name.
Ki fuji	Wistaria chinensis	Yellow wistaria	Tampopo	Taraxacum officinale	
Fuji-matsu	Larix leptolepis		Sumire	Viola patrini	Violet

4th MONTH (PRESENT MAY).

Japanese Name.	Botanical Name.	English popular Name.	Japanese Name.	Botanical Name.	English popular Name.
Kiku	Chrysanthemum coronarium	Chrysanthemum	‡Enishida	Cytisus scoparius	
Kuchinashi	Gardenia florida		‡Hama-nadeshiko	Dianthus japonicus	
Sakaki	Cleyera japonica		‡Tsurigane so	Campanula punctata	Bluebell
Azami	Cnicus japonicus		*Hiyo riu	Hypericum chinense	
Tochi-so	Clintonia udensis		‡Kodemari	Spiræa cantoniensis	
Akaza	Chenopodium album		‡Suwo	Cæsalpinia Sappon	
*Kakitsubata	Iris lævigata	Iris	‡Kobushi	Magnolia kobus	Magnolia
*Botan	Pæonia moutan	Tree Peony	*Shoga	Iris japonica	Iris
‡Ayame	Iris sibirica	Iris	‡Mokuren	Magnolia conspicua	Magnolia
‡Ippatsu	Iris tectorum	Iris	*Shimotsuke	Spiræa japonica	
*Omoto	Rhodea japonica		*Sendai hagi	Thermopsis fabacea	
*Kusa-aoi	Althæa rosea		‡Shiran	Bletia hyacinthina	
*Shakuyaku	Pæonia albiflora	Peony	*Teppo yuri	Lilium longiflorum	Lily
Utsugi	Deutzia Sieboldiana		*Sasa yuri	Lilium japonicum	Lily
*Yuri	Lilium	Lily	*Hime yuri	Lilium concolor	Lily
‡Zakuro	Punica granatum		‡Suzu-ran	Convallaria majalis	
*Sendan	Melia azedarach		*Hankwai so	Senecio japonicus	
Tsutsuji	Rhododendron indicum	Azalea	*Giboshi	Funkia ovata	
*Hana-nanten	Nandina domestica		*Kumagaye-so	Cypripedium japonicum	
*Kirishima	Rhododendron obtusum		Atsumori so	Cypripedium macanthrum	
‡Kinsenkwa	Calendula officinalis		*Benkei so	Sedum erythrostictum	
‡Kohone	Nuphar japonicum		*Hana-yu	Citrus aurantium	
*Tessen	Clematis florida	Clematis	*Kikoku	Citrus fusca	
*Natsu-yuki	Deutzia Sieboldiana		*Shuro	Chamærops excelsa	
‡Kirin-so	Sedum kamtschaticum		*Koboku	Magnolia hypoleuca	Magnolia
*Eijin-so	Papaver rhœas		*Tampopo	Taraxacum officinale	
‡Futo-i	Scirpus lacustris		*Fuji-kazura		

5th MONTH (PRESENT JUNE).

Japanese Name.	Botanical Name.	English popular Name.	Japanese Name.	Botanical Name.	English popular Name.
*Kiku	Chrysanthemum coronarium	Chrysanthemum	*Ajisai	Hydrangea hortensis	Hydrangea
*Giboshi	Funkia ovata		*Tochi so	Clintonia udensis	
‡Shiran	Bletia hyacinthina		*Shimotsuke	Spiræa japonica	
*Kuchinashi	Gardenia florida		*Natsu-yuki	Deutzia Sieboldiana	
*Uki-kusa	Lemna minor		*Mokko kwa	Rosa Banksiæ	

Japanese Name.	Botanical Name.	English popular Name.	Japanese Name.	Botanical Name.	English popular Name.
‡Hime-yuri	Lilium concolor	Lily	*Sakaki	Cleyera japonica	
*Sen-no	Lychnis senno		‡Kodemari	Spiræa cantoniensis	
*Matatabi	Actinidia polygama		*Kwaku so	Phajus grandiflorus	
*Zakuro	Punica granatum		*Futo-i	Scirpus lacustris	
‡Biyo-riu	Hypericum chinense		‡Hoso i	Juncus communis	
‡Hana-nanten	Nandina domestica		‡Sankaku-i	Scirpus lacustris	
‡Tessen	Clematis florida	Clematis	*Kayatsuri-gusa	Cyperus iris	
‡Kiri-shima	Rhododendron obtusum	Azalea	*Sendan	Melia azedarach	
*Satsuki	Rhododendron macranthum	Azalea	*Hana shobu	Iris lævigata	Flag
*Kirin-so	Sedum kamtschaticum		*Kusa ayame	Iris sibirica	Wild Iris
*Natsu-zukashi	Lilium Thunbergianum	Lily	*Mankaishi	Vitex trifolia	
*Kingin-kwa	Goodyera parviflora		*Nichi-nichi so	Vinca rosea	
*Nadeshiko	Dianthus superbus		*Kokwa	Carthamus tinctorius	
*Kawara-nadeshiko	Dianthus superbus		‡Omoto	Rhodea japonica	
‡Teppo-yuri	Lilium longiflorum	Lily	‡Kakitsubata	Iris lævigata	Iris
‡Sasa yuri	Lilium japonicum	Lily	*Hakucho-ke	Serissa fœtida	
‡Hankwai-so	Senecio japonicus		*Kwannon-so	Reineckia carnea	
‡Benkei so	Sedum erythrosticum		*Kurumi	Juglans regia	
*Kuma-yanagi	Berchemia racemosa		*Ochi	Melia japonica	
*Kohone	Nuphar japonicum		‡Kobotan	Magnolia hypoleuca	Magnolia

6th MONTH (PRESENT JULY).

Japanese Name.	Botanical Name.	English popular Name.	Japanese Name.	Botanical Name.	English popular Name.
‡Kiku	Chrysanthemum coronarium	Chrysanthemum	*Mizu-aoi	Monochoria vaginalis	
‡Hana-nanten	Nandina domestica		*Oguruma	Inula britannica	
‡Omoto	Rhodea japonica		*Mokuge	Hibiscus syriacus	
*Ran		Orchid	*Nozen kwa	Tecoma grandiflora	
‡Oshiroi-bana	Mirabilis jalapa		‡Kirin so	Sedum kamtschaticum	
*Seki chiku	Dianthus chinensis	Kind of Ilium! co	‡Benkei-so	Sedum erythrosticum	
*Hishi	Trapa bispinosa		‡Medo hagi	Lespedeza sericea	
‡Hakucho ke	Serissa fœtida		*Asagao	Ipomæa hederacea	Morning Glory
*Nadeshiko	Dianthus superbus		*Hiru gao	Convolvulus japonicus	Convolvulus
*Hasu	Nelumbium speciosum	Lotus	*Yu-gao	Nuphar japonicum	Convolvulus
‡Hi-ogi	Pardanthus chinensis		*Kohone	Nuphar japonicum	
‡Giboshi	Funkia ovata		*Kakitsubata	Iris lævigata	Iris
*Kuzu	Pueraria Thunbergiana		*Kawara-nadeshiko	Dianthus superbus	
*Tora no o	Lysimachia clethroides		‡Futo i	Scirpus lacustris	
*Kikio	Platycodon grandiflorum		‡Hoso i	Juncus communis	
*Tsuta	Vitis incomstans	Ivy	‡Sankaku i	Scirpus lacustris	
*Gampi	Lychnis grandiflora		*Anju	Dianthus Caryophyllus	
‡Senno	Lychnis senno		‡Shiu kaido	Begonia evansiana	
*Karaguruma	Clematis patens	Clematis	‡Otogiri-so	Hypericum erectum	
*Sakaki	Cleyera japonica				

Japanese Name.	Botanical Name.	English Popular Name.	Japanese Name.	Botanical Name.	English Popular Name.
Omodaka	Alisma plantago		Natsu-tsubaki	Stuartia pseudo-camellia	
Zakuro	Punica granatum		O-yuri	Lilium	Lily
Sanzashi	Crataegus cuneata		Sasa-yuri	Lilium japonicum	Lily
Manjusake	Nerine japonica		Teppo-yuri	Lilium longiflorum	Lily
Sendan	Melia azedarach		Hime-yuri	Lilium concolor	Lily
Mankeishii	Vitex trifolia		Natsu-zukashi	Lilium Thunbergianum	Lily
Kwannon-so			Itadori	Polygonum cuspidatum	

7th MONTH (PRESENT AUGUST).

Japanese Name.	Botanical Name.	English Popular Name.	Japanese Name.	Botanical Name.	English Popular Name.
Kiku	Chrysanthemum coronarium	Chrysanthemum	Otogiri-so	Hypericum erectum	
Kikio	Platycodon grandiflora		Shion	Aster tartaricus	Aster
Ran		Orchid	Keito	Celosia argentea	
Gampi	Lychnis grandiflora		Sawa-gikio	Lobelia sessilifolia	
Mokuge	Hibiscus syriacus		Hosen-kwa	Impatiens balsamina	
Tsuta	Vitis inconstans	Ivy	Fuyo	Hibiscus mutabilis	
Sennichi-so	Gomphrena globosa		Ha-geito	Amaranthus melancholicus	
Medo-hagi	Lespedeza sericea		Dandoku	Canna indica	
Hasu	Nelumbium speciosum	Lotus	Hi-ogi	Pardanthus chinensis	
Oguruma	Inula britannica		U-kon	Curcuma longa	
Senrio	Chloranthus brachystachys		Kichijo-so	Reineckia carnea	
Kuzu	Pueraria Thunbergiana		Kushide	Rhus semi-alata	
Ominayeshi	Patrinia scabiosæfolia		Kakitsubata	Iris lævigata	Iris
Asagao	Ipomœa hederacea		Tori-kabuto	Aconitum Fischeri	
Hishi	Trapa bispinosa		Manjusake	Lycoris radiata.	
Yu-gao		Convolvulus	Mankeishi	Vitex trifolia	
Hiru-gao	Convolvulus japonicus	Convolvulus	Benkei-so	Sedum erythrostictum	
Hagi	Lespedeza bicolor		Hakuchoke	Serissa fœtida	
Shiu-kaido	Begonia Evansiana		Kwannon-so	Reineckia carnea	
Kohone	Nuphar japonicum		Riukiu-giku		Chrysanthemum
Futo-i	Scirpus lacustris		Aoi	Althæa rosea	
Hoso-i	Juncus communis		Tsuru-modoki	Celastrus articulatus	
Sankaku-i	Scirpus lacustris		Anja	Dianthus caryophyllus	
Mizu-noi	Monochoria vaginalis		Nadeshiko	Dianthus superbus	
Omodaka	Alisma Plantago		Kawara-nade-shiko	Dianthus superbus	

8th MONTH (PRESENT SEPTEMBER).

Japanese Name.	Botanical Name.	English Popular Name.	Japanese Name.	Botanical Name.	English Popular Name.
Kiku	Chrysanthemum coronarium	Chrysanthemum	Tsuta	Vitis inconstans	Ivy
Susuki	Eularia japonica		Hagi	Lespedeza bicolor	
Hasu	Nelumbium speciosum		Ogi		Kind of Reed

Japanese Name.	Botanical Name.	English Popular Name.	Japanese Name.	Botanical Name.	English Popular Name.
Kakitsubata	Iris laevigata	Iris	Hassaku-bai		Autumn Plum
Shion	Aster tataricus	Aster	Usu-momiji	Acer palmatum	Kind of Maple
Yukinoshita	Saxifraga sarmentosa		Sanzashi	Crataegus cuneata	
Fujibakama	Eupatorium chinense		Hama-giku	Chrysanthemum nipponicum	
No-giku		Wild Chrysanthemum	Waremoko	Poterium officinale	
			Okina-gusa	Anemone cernua	
Hosen-kwa	Impatiens balsamina		Medo-hagi	Lespedeza sericea	
Fuyo	Hibiscus mutabilis		Sawa-gikio	Lobelia sessilifolia	
Hi-mawari	Helianthus annuus	Sunflower	Sennichi so	Gomphrena globosa	
Keito	Celosia argentea		Mizu-aoi	Monochoria vaginalis	
Ominayeshi	Patrinia scabiosaefolia		Kohone	Nuphar japonicum	
Otokoyeshi	Patrinia scabiosaefolia alba		Nishiki-bana	Euonymus alatus	
Tori-kabuto	Aconitum Fischeri		Benkei so	Sedum erythrosticum	
Tsuru-modoki	Celastrus articulatus		Kwannon so		
Ume-modoki	Ilex bieboldi		Ukon	Curcuma longa	
Mokuge	Hibiscus syriacus		Kichijo so	Reineckia carnea	
Gan-rai ko	Amaranthus melancholicus		Kushide	Rhus semi-alata	
Karukaya	Anthistiria arguens		Riukiu-giku		Chrysanthemum
Rindo	Gentiana scabra		Kakitsubata	Iris laevigata	Iris
Uzura-gusa		Kind of Grass	Nishikigi	Euonymus alatus	

9th MONTH (PRESENT OCTOBER).

Japanese Name.	Botanical Name.	English Popular Name.	Japanese Name.	Botanical Name.	English Popular Name.
Kiku	Chrysanthemum coronarium	Chrysanthemum	Hama-giku	Chrysanthemum nipponica	Chrysanthemum
Nanten	Nandina domestica		Sanzashi	Crataegus cuneata	
Omoto	Rhodea japonica		Ominayeshi	Patrinia scabiosaefolia	
Hagi	Lespedeza bicolor		Otokoyeshi	Patrinia scabiosaefolia alba	
Ume modoki	Ilex Sieboldi		Kocho ke		
Tsuru-modoki	Celastrus articulatus		Shikizaki	General term for flowers blooming in all four seasons	
Ogi		Kind of Reed	Kabuto-giku	Aconitum Fischeri	
Rindo	Gentiana scabra		Mizuhiki	Polygonum filiforme	
Suisen	Narcissus tazetta	Narcissus	Fujibakama'	Eupatorium chinense	
Susuki	Eulalia japonica		Yuki-no shita	Saxifraga sarmentosa	
Sawa-gikio	Lobelia sessilifolia		Waremoko	Poterium officinale	
Tsuwa buki	Senecio Kampferi		Medo hagi	Lespedeza sericea	
Cha-no-hana	Camellia thuifera	Tea Plant	No-giku		Wild Chrysanthemum
Yatsude	Fatsia japonica				
Sazankwa	Camellia sasanqua	Camellia	Uzura-gusa		Kind of Grass
Tsuta	Vitis inconstans	Ivy	Riukiu-giku		Chrysanthemum
Biwa	Photinia japonica		Kohone	Nuphar japonicum	
Shion	Aster tataricus	Aster	Koyo-moto	General term for trees the leaves of which redden in the Autumn	
Kakitsubata	Iris laevigata	Iris	Yanagi-no-rui	Trees of the Willow kind	
Karukaya	Anthistiria arguens		Nishikigi	Euonymus alatus	

10th MONTH (PRESENT NOVEMBER).

Japanese Name.	Botanical Name.	English Popular Name.	Japanese Name.	Botanical Name.	English Popular Name.
‡Zan-giku		Late Chrysan-themum	*Shikizaki	General term for flowers blooming in all four seasons	
*Suisen	Narcissus tazetta	Narcissus	*Koyo mono	General term for trees the leaves of which turn red in the Autumn	
*Kan-giku	Pyrethrum sinense	Winter Chrysan-themum	*Nebuka so	Allium fistulosum	
‡Sanzashi	Cratægus cuneata		*Yuki no shita	Saxifraga sarmentosa	
*Cha no hana	Camellia thefera	Tea Plant	*Yatsu de	Fatsia japonica	
*Biwa	Photinia japonica		*Karukaya	Anthistiria arguens	
‡Nanten	Nandina domestica		‡Rindo	Gentiana scabra	
*Omoto	Rhodea japonica		‡Hiyozashi-Tsubaki	Camellia japonica	Early Camellia
*Neko-yanagi	Salix brachystachys	Kind of Willow	‡Toji-bai	Prunus mume	Early Plum
‡Tsuwa buki	Senecio Kæmpferi		*Jugwatsu zakura	Prunus pseudo-cerasus	Tenth Month Cherry
Kocho ke					

11th MONTH (PRESENT DECEMBER).

Japanese Name.	Botanical Name.	English Popular Name.	Japanese Name.	Botanical Name.	English Popular Name.
*Kan-giku		Winter Chrysan-themum	*Koyo mono	General term for trees of which turn red in the Autumn	
*Suisen	Narcissus tazetta	Narcissus	‡Kocho ke		
*Nanten	Nandina domestica		‡Kan-botan	Pæonia moutan	Winter Peony
‡Omoto	Rhodea japonica		‡Sazankwa	Camellia Sasanqua	Camellia
*Neko-yanagi	Salix brachystachys		*Yatsu de	Fatsia japonica	
‡Toji-bai	Prunus mume	Early Plum	‡Tsubaki	Camellia japonica	Camellia
‡Jugwatsu zakura	Prunus pseudo-cerasus	Tenth Month Cherry	*Shikizaki	General term for flowers blooming in all four seasons	
‡Biwa	Photinia japonica				

12th MONTH (PRESENT JANUARY).

Japanese Name.	Botanical Name.	English Popular Name.	Japanese Name.	Botanical Name.	English Popular Name.
‡Kangiku		Winter Chrysan-themum	‡Yanagi-no-rui	Various kinds of Willow	
‡Suisen	Narcissus tazetta	Narcissus	*Ro bai	Chimonanthus fragrans	
*Kan-botan	Pæonia moutan	Winter Peony	‡Kinsenkwa	Calendula officinalis	
‡Nanten	Nandina domestica		‡Kocho-ke		
‡Omoto	Rhodea japonica		*Koyo-no-rui	Various trees the leaves of which redden in the Autumn	
‡Murozaki momo		Forced Peach	‡Shikizaki	Flowers blooming in all four seasons	
‡Haku-bai	Prunus mume flore alba		‡Rengio	Forsythia suspensa	
*Tsubaki	Camellia japonica	Camellia			

Against the Japanese names in the above list of flowers certain signs are placed to indicate the following distinctions :—

* Stands for those trees or plants which are termed *Living Flowers*, being particularly characteristic of the month under which they are placed. Such flowers are much prized for felicitous occasions.

† Distinguishes the *Early Flowerings*,— flowers which are in advance of their proper season in the month under which they are placed. These have also their appropriate use in floral compositions.

‡ Indicates what are called *Passed Flowers*, and ‖ stands for what are termed *Dead Flowers*. These names have reference to flowers which are late or passed in month or season, belonging properly to earlier months. The use of such flowers is forbidden for most ceremonial occasions. There exists also the term *Vulgar Flowers*, applied to wild plants, or to those of very common character which possess no fancy name ; and the employment of flowers included under this head is not permitted except in the hands of the most experienced professors of the art. The use of cereals is also to be avoided.

Considerations of good or evil luck enter largely into the choice of flowers, for floral arrangements, especially when employed as decorations for occasions of rejoicing; and there are certain flowering plants and trees reputed to possess poisonous properties in their roots, stems, leaves, or blossoms, the use of which is objected to at any time, their employment being considered unlucky and ominous. The following is a list of the principal of such poisonous flowers :—

OMINOUS FLOWERS.

Japanese Name.	Botanical Name.	Note.	Japanese Name.	Botanical Name.	Note.
Mochi-tsutsuji	Rhododendron ledifolium	The white flower species is not poisonous	Asebo	Andromeda japonica	Stem poisonous
Yaye-kwanzo	Hemerocallis fulva	Single flower species not poisonous	Yamagobo	Rhaponticum atriplicifolium	Red kind poisonous
Manjuzake	Nerine japonica	Leaves are poisonous	Tori-kabuto	Aconitum Fischeri	Root poisonous
Hana-suwo	Cercis chinensis	Flower poisonous	Karasu-ogi	Rumex aquaticus	Stem said to be poisonous
Goji-kwa	Pentapetes phœnicea	Highly poisonous	Inu-kusu	Machilus Thunbergii	Root poisonous
Nozen-kazura	Tecoma grandiflora	Tendrils poisonous	Tachimachi-gusa	Aconitum lycoctonum	Very poisonous
Yama-ajisai	Hydrangea hirta	Root poisonous	Gibo-hi	Funkia ovata	Flower poisonous
Hosen-kwa	Impatiens balsamina	Leaves poisonous	Yama-nasubi	Datura alba	Very poisonous
Miyama-shikimi	Skimmia japonica	Leaves poisonous	Konnïaku	Arisæma japonica	Root poisonous
Hito-koshi		Strongly poisonous	Kusagi	Clerodendron trichotomum	Leaves poisonous
Yatsu-de	Fatsia japonica	Root poisonous			

In addition to the last named, all flowers having a strong odour are considered unsuitable for placing before guests.

Among the flowers peculiar to the different months, as previously classified, some are considered specially appropriate for displaying upon felicitous occasions, whilst others, though used at ordinary times, are interdicted for such important occasions.

As most of these flowers are to be found enumerated in the complete tables already given, the following classification is abbreviated, merely giving the Japanese names and the corresponding name in English, botanical names being printed only where no English equivalent exists. It may be observed that this list includes in all twenty-four species of plants and trees, and if different species of the same *genera* be classed together, the number of specially honoured flowers becomes reduced to fifteen :—

FLOWERS SUITABLE FOR FELICITOUS OCCASIONS.

1st MONTH (PRESENT FEBRUARY).

Fukuju-so	*Adonis amurensis*
Yabu-koji	*Ardisia japonica*
Haku-bai	White Plum
Yanagi	Willow
Omoto	*Rhodea japonica*
Choshun	*Rosa indica*
Shu-chiku-bai	Combination of Pine, Bamboo, and Plum.

2nd MONTH (PRESENT MARCH).

Momo	Peach
Yanagi	Willow
Ko-bai	Red Plum
Omoto	*Rhodea japonica*
Haru-giku	Spring Chrysanthemum
Choshun	*Rosa indica*

3rd MONTH (PRESENT APRIL).

Sakura	Cherry
Momo	Peach
Haru-giku	Spring Chrysanthemum
Omoto	*Rhodea japonica*
Choshun	*Rosa indica*

4th MONTH (PRESENT MAY).

Botan	Tree Peony
Shakuyaku	Peony
Moso-chiku	Bamboo
Omoto	*Rhodea japonica*
Choshun	*Rosa indica*
Kiku	Chrysanthemum

5th MONTH (PRESENT JUNE).

Kiku	Chrysanthemum
Omoto	*Rhodea japonica*
Moso-chiku	Bamboo
Choshun	*Rosa indica*

6th MONTH (PRESENT JULY).

Kiku	Chrysanthemum
Omoto	*Rhodea japonica*
Nanten	*Nandina domestica*
Choshun	*Rosa indica*

7th MONTH (PRESENT AUGUST).

Kiku	Chrysanthemum
Omoto	*Rhodea japonica*
Choshun	*Rosa indica*
Nanten	*Nandina domestica*

8th MONTH (PRESENT SEPTEMBR).

Kiku	Chrysanthemum
Omoto	*Rhodea japonica*
Hassaku-bai	Autumn Plum
Nanten	*Nandina domestica*
Choshun	*Rosa indica*

9th MONTH (PRESENT OCTOBER).

Kiku	Chrysanthemum
Omoto	*Rhodea japonica*
Nanten	*Nandina domestica*
Choshun	*Rosa indica*
Suisen	Narcissus
Yanagi	Willow

10th MONTH (PRESENT NOVEMBER).

Zan-giku	Chrysanthemum
Suisen	Narcissus
Omoto	*Rhodea japonica*
Nanten	*Nandina domestica*
Choshun	*Rosa indica*
Yanagi	Willow

11th MONTH (PRESENT DECEMBER).

Suisen	Narcissus
Kan-giku	Winter Chrysanthemum
Omoto	*Rhodea japonica*
Yanagi	Willow
Nanten	*Nandina domestica*
Toji-bai	Early Plum
Choshun	*Rosa indica*
Yaye-tsubaki	Double Camellia

12th MONTH (PRESENT JANUARY).

Suisen	Narcissus
Kan-giku	Winter Chrysanthemum
Yanagi	Willow
Omoto	*Rhodea japonica*
Haku-bai	White Plum
Murozaki-momo	Forced Peach
Choshun	*Rosa indica*
Tsubaki	Camellia

The following list of flowers, the use of which is prohibited for special occasions of ceremony or congratulation, is arranged without regard to the months to which they belong. The employment of such flowers is deprecated at any season, without reference to any particular month, unless no other flowers can possibly be obtained. The reasons for their rejection are not always clear; sometimes the objection is to the form, sometimes to the colour, occasionally to some supposed poisonous property, and often to what seems little more than traditional superstition or caprice. It is not surprising therefore to find these rules disregarded at times by certain masters :—

FLOWERS PROHIBITED FOR FELICITOUS OCCASIONS.

Ito-susuki	*Eularia japonica*	Nashi	*Pyrus ussuriensis*
Shion	*Aster*	Kanzo	*Hemerocallis flava*
Basho	*Musa Banjoo*	Fuyo	*Hibiscus mutabilis*
Shinobu	*Davallia bullata*	Renge	*Lotus*
Kawara-nadeshiko	*Dianthus superbus*	Hototogisu-so	*Tricyrtis japonica*
Ogi	*Kind of Reed*	Mokuren	*Magnolia*
Hagi	*Lespedeza bicolor*	Cha-no hana	*Camellia theifera*
Fujibakama	*Eupatorium chinense*	Ran	*Orchid*
Tsutsuji	*Azalea*	Dandoku	*Canna indica*
Niwatoko	*Sambucus racemosa*	Rengio	*Forsythia suspensa*
Miyama-shikimi	*Skimmia japonica*	Yoshi	*Phragmitis communis*
Himuro	*Thuya squarrosa*	Ashi	*Phragmitis communis*
Karukaya	*Chinese Bamboo*	Rindo	*Gentiana scabra*
Keito	*Celosia argentea*	Awayuki	*Gnaphalium Sieboldiianum*
Hiro-gashiwa	*Large-leafed Oak*	Shakunage	*Rhododendron*
Karukaya	*Anthistiria arguens*	Kuchi-nashi	*Gardenia florida*
Jincho-ke	*Daphne odora*	Asa-gao	*Ipomoea holeracea*
Hashibami	*Corylus heterophylla*	Gibashi	*Funkia ovata*
Hiaku-jikko	*Lagerstroemia indica*	Iti-ngi	*Pardanthus chinensis*
Zakuro	*Punica granatum*	Ume-modoki	*Ilex Sieboldi*
Mokuge	*Hibiscus syriacus*	Yama-nashi	*Smilax biflora*
Mitsu-mata	*Edgeworthia papyrifera*	Kohone	*Nuphar japonicum*
Hosen-kwa	*Impatiens balsamina*	Hinoki	*Thuya obtusa*
Koikwa	*Carthamus tinctorius*	Yatsude	*Fatsia japonica*
Kushi	*Papaver somniferum*	Ajisai	*Hydrangea hortensis*

Ranking highest in the above list of felicitous flowers, the following seven are considered as *par excellence* those for ceremonies and congratulatory occasions :—

The *Kiku*, or Chrysanthemum, to which is given the fancy name *Choju-so*, meaning *Long-lasting Plant*, on account of its growing through all the four seasons.

The *Suisen*, or Narcissus, called by the fancy name of *Inyo-so*, or *Plant of the Two Sexes*, because it comes in the Winter and lasts till the Spring of the following year.

The *Momiji*, or Maple, fancifully called *Dokuge-so* or *Poison-dispelling Plant*, because there is an idea that it absorbs all poison and infection from the air.

The *Sakura*, or Cherry, regarded in Japan as the king of flowers.

The *Botan*, or Tree Peony, fancifully named *Fuki-gusa*, meaning *Plant of Wealth and High Rank*. In China the Peony is said to be the queen of flowers.

The *Omoto*, or Rhodea japonica, much honoured because, unaffected by heat or cold, its leaf remains strong and green throughout the year.

The *Fuji*, or Wistaria, fancifully called *Niki-so*, meaning *Plant of the Two Seasons*, because, appearing between the third and fourth months, it belongs both to Spring and Summer. Though much honoured and used for felicitous occasions, the Wistaria must not be employed at weddings on account of its purple colour.

In addition to the above seven flowers, the *Kakitsubata* (Iris lævigata) also takes high rank, but on account of its purple colour is, like the Wistaria, prohibited for wedding ceremonies.

Hitherto attention has been directed to the principal flowering plants and trees of the country, and to the degree of esteem in which they are individually regarded, especially with reference to particular months. But many floral compositions consist of combinations of two or more different kinds of growth. The manner in which different species are thus combined is best explained when the whole theory of the Japanese methods of arrangement is discussed in a later chapter. But apart from the manner of grouping them, there are certain prejudices in favour of and against different combinations of material which require mention in the present context. The following is a list of a few of the suitable and unsuitable combinations of flowers : —

APPROPRIATE COMBINATIONS.

Matsu (Pine)	with *Cho-kan* (Rosa indica)	*Lowano-ichi* (Ilex Sicboldi)	with *Suisen* (Narcissus)
Matsu (Pine)	with *Kiku* (Chrysanthemum)	*Haran* (Orchid)	with *Nadeshiko* (Dianthus superbus)
Yanagi (Willow)	with *Suisen* (Narcissus)		
Momiji (Maple)	with *Kiku* Chrysanthemum, white or yellow)	*Tokusa* (Equisetum hyemale)	with *Senno* (Lychnis senno)
Tsubaki (Camellia)	with *Suisen* (Narcissus)	*Haku-bai* (White Plum)	with *Kinsenka* (Calendula officinalis)

Momo (Peach)	with Yamabuki (Kerria japonica)	Take (Bamboo)	with Asagao (Morning Glory)
Futoi (Scirpus lacustris)	with Kakitsubata (Iris laevigata)	Nanten (Nandina domestica)	with Shira-giku (White Chrysanthemum) or Suisen (Narcissus)
Kohone (Nuphar japonicum) or			

OBJECTIONABLE COMBINATIONS.

Kashiwa (Oak)	with Shion (Aster)		Maki (Podocarpus macrophylla)
Take (Bamboo)	with Susuki (Eularia japonica)		
Ume (Plum)	with Yanagi (Willow)	Momo (Peach)	with Sakura (Cherry)
Nanten (Nandina domestica)	with Take (Bamboo)	Maki (Podocarpus macrophylla)	with Haran (Orchid)
Enoki (Celtis sinensis)	with Nadeshiko (Dianthus superbus)	Hinoki (Thuya obtusa)	with Haran (Orchid)
Tsubaki (Camellia)	with Kinsenkwa (Calendula officinalis)	Shaga (Iris japonica)	with Haran (Orchid)
Matsu (Pine)	with Zakuro (Punica Granatum) or		Omoto (Rhodea japonica) or
			Kohone (Nuphar japonicum)

All of the above combinations, good and bad, are of trees or plants which are in bloom during the same month, and are therefore arrangements which are practicable without violating the rules as to seasonableness. Those which are classified as objectionable are so considered therefore for other reasons, which are various. Sometimes too close a resemblance in form or colour is the reason of the objection; in other cases, similarity of species, or of locality of production, leading to redundancy of expression in the composition, is the cause of the dislike. The Peach and the Cherry, for example, being both flowering trees and somewhat similar in character, are not considered suitable in combination.

HISTORY AND THEORY.

BEFORE proceeding further with an explanation of the Floral Art, it may be of some interest to enquire into the origin of a cult so curiously unlike, in its methods, to any hitherto followed in Europe. Japanese historians claim for it an Indian and religious origin. The same Buddhist doctrine which deprecated the wanton sacrifice of animal life is said to have suggested the gathering of flowers, liable to rapid destruction in a tropical climate, and prolonging their life by careful preservation. The survival of such a theory would seem to show that some form of the art was first introduced into this country with the adoption of the Buddhist faith, both as a part of its ritual (flowers being placed before the Buddhist spirits), and also as forming a pious pastime for the priests. The religion of Buddha, as is well known, reached Japan through China in the sixth century, and certain Chinese priests are referred to as the first teachers of the art of arranging flowers in Japan. It also appears that the earliest native practitioners in this country were famous priests, amongst whom Shotoku Taishi and Meikei Shonin are particularly mentioned.

These primitive flower compositions were, however, of a more accidental and, at the same time, of a more extravagant character than those of the art as it became afterwards modified and developed. They partook more of the nature of a Western bunch or nosegay, being crowded in arrangement and miscellaneous in substance, and without the severe conventionality of later methods. The style of composition adopted still survives under the name of *Shia-no-hana*, meaning Central Flower Arrangement, and is used for flowers placed as religious ornaments or offerings before shrines and tombs. An approach to symmetry was a governing feature of such compositions, the designers not having yet developed that less formal kind of balance discoverable in the modern style which is more in harmony with nature and admits of endless variety of form. Branches of blossoming trees or foliage were employed in their natural state, without artificial bending or trimming, to form a vertical central mass; and other flowers or bunches of foliage were disposed on either side in balancing groups. The idea of imparting graceful and harmonious curves to the different lines of the composition was as yet undeveloped. Unlike the later and more refined flower arrangements, this early style was remarkable for the mixture of a variety of different materials. The lines of a composition, distinguished by such terms as *centre*, *sub-centre*, and *support*, were respectively formed of branches of different growth, some of which were in full leaf or flower, and others purposely light and sparse in character. Large leaves of other plants were used at the base or connection of these various branches, to hide their bareness below, and careful attention was given to the bends and curves of these leaves so as to reveal their front and back surfaces in a well balanced contrast. Even in this comparatively early form of the art, the proportion which the floral composition held to the vessel which contained it, was fixed by rule, a practice which was followed in all later arrangements. Various examples of the Central Flower Arrangement are shown in Plates VII., VIII., and IX.

Another early style of arrangement existed, suited to broad basin-like vessels, and adapted to the employment of large stumps of trees in divided groups. Such compositions revealed no attempt at symmetry, but resembled somewhat a kind of miniature gardening. Heavy branched trunks of willow, pine, plum, maple, or other favourite trees were grouped with divided spaces between, and sometimes with plants or grasses, added at their base. An example of this style of arrangement may be seen in Plate X.

So close was the connection between flower vessels and flower arrangements, that the more modern development of the Floral Art was simultaneous with a great fancy for ornamental vases of various kinds which sprang up under the patronage of the famous Regent Yoshimasa, at Higashi-yama. The impetus given to the manufacture of choice vessels by this Prince, chiefly owing to his inauguration of the Tea Ceremonies and other polite accomplishments, was accompanied by an austere refinement in the methods of arranging flowers. It was mainly with the object of contributing to the Tea Ceremonial that the first modifications in the Flower Art took place, and the chief reformers were the *Chajin*, or Professors of Tea. Sen no Rikiu, Senke, Sekishiu, and Enshiu, who were four of the principal flower designers of these and subsequent times, were among the most famous professors of the Tea Ceremonial. But the affected simplicity which ruled the Tea Room did not give full scope for the exercise of great elaboration in flower compositions, and other less austere forms of arrangement were developed, suited to the larger chambers of the nobility, but based upon the principles which had thus been inaugurated. Flower compositions, as finally classified, are given different character and proportion according to the special class of chamber which they adorn, or the particular rank of the person in whose residence they are used.

The arrangement of flowers has always been regarded in Japan as an occupation befitting learned men and *literati*. Ladies of the aristocracy have practised it, as they have other arts, but it is by no means considered as an effeminate accomplishment. Priests, philosophers, and men of rank who have retired from public life have been its most enthusiastic followers.

Mixed up with the theory of the art, and imparting to it at first sight an air of quaintness and mystery, is a considerable amount of Chinese philosophy, together with many traditional superstitions. Ideas of good and evil luck control both the selection of material and the manner of arrangement for special occasions. Various virtues are attributed to professors of the art, who are considered to belong to a sort of aristocracy of talent, enjoying privileges of rank and precedence in society to which they are not by birth entitled. A religious spirit, self denial, gentleness, and forgetfulness of cares, are some of the virtues said to follow from a habitual practice of the art of arrangement flowers. Philosophical classifications are resorted to for the purpose of distinguishing the different parts of floral designs. Thus Earth, Heaven, and Mankind are names given by some Schools to the members of a tri-lineal flower arrangement; Earth, Fire, Water, Metal, and Wood being used in the same way to designate the parts of a five-lined design. Other Schools apply, in a similar way, the names of abstract ideas, such as those of the five orders of Japanese versification, or the virtues of the human heart. Such different methods of nomenclature are numerous, and would appear to have been adopted by the different Schools principally to impart an appearance of originality and mystery to their own particular versions of one and the same art.

Again, the male and female principles of Chinese philosophy are constantly applied to distinguish contrasting forms, surfaces, or colours in composition. It has always been a favourite fancy of the Japanese to apply distinctions of sex to inanimate nature. In landscape, and landscape-gardening, it is customary to discriminate between *male* and *female* cascades, *male* and *female* plants and trees, and *male* and *female* rocks and stones. The distinction is not so much one of individual form as of forms placed in combination or contrast, which are regarded as *male* or *female* in respect of one another. Thus the main torrent of a waterfall is considered *masculine*, and the lower fall in proximity is regarded as *feminine*. In like manner, rocks used in gardening have no distinguishing sex, unless they are used in pairs or groups. In the case of two stones of different character placed side by side, the one of bolder and more vigorous shape will be called the *male*, and the other will be called the *female* stone. Curious as such fancies may seem, they are of considerable value when applied to the art of design, their observance helping to produce that harmony of well balanced contrasts which should pervade all compositions. Nor are such ideas, indeed, quite foreign to Western art, in which, for example, it is common to attribute *male* and *female* characteristics to the different orders of architecture.

In the Floral Art the idea of sex is applied to contrasting forms, as of long and short, large and small, angular and round, as well as to different kinds of growth and to various colours in combination. When a flowering tree is used in combination with a plant, the tree is considered as *male* and the plant as *female*. With flowers, buds are regarded as *female*, full flowers as *male*, and overblown flowers again are classed as *female*; the time of full vigour receives the *male*, and the periods of weakness the *female* character. A similar fancy is applied to the different surfaces of leaves and to the different colours of flowers. Among colours, Red, Purple, Pink, and variegated colours are *male*; and Blue, Yellow, and White are *female*. The front surface of a leaf is *male*, and its under surface is *female*. With flowers, the different forms of bud and full blossom, and with leaves, the different surfaces lend themselves easily to such distinctions; but berries seem to have defeated attempts at sexual classification, until Enshiu, one of the masters of the Flower Art, after observing the bulbul bird pecking at the fruit of a tree, devised the method of marking certain of the berries in a floral composition as if pecked by birds, and thus imparting to them a dual character,—that of square and round. The *male* and *female* principles are also applied to the different directions of the stems of a composition, the right hand side and front of the design being regarded as *male*, and the left hand and further side as *female*. A stem on the left side of a composition turned to the front and back to the right is said to have *male* character, while one on the right bent back to the left is said to have *female* character.

In addition to sex, the different colours in flowers or leaves are said to have order of *rank*. This idea of rank is applied principally to different colours of the same flower. In most cases the white flower of every species takes highest rank, but there are exceptions to this rule. The yellow Chrysanthemum takes precedence of those of any other colour. With Peach blossoms, the pale pink colour ranks first, though a rich red specimen exists. In the case of other flowers the colours taking the highest rank are with Irises, purple; with Camellias, red; with Wistarias, lavender; with Peonies, red; with Valerians, yellow; with Convolvuli, dark blue; with Kerrias, yellow; with Kikios, light purple; and with Lespedezas, pink.

As before mentioned, a number of different Schools of Flower Arrangement sprang up from the time of the great art-patron, Yoshimasa. The styles of composition practised by these different Schools are known as the *Ko* style, *Enshiu* style, *Shiuslo* style, *Schishiu* style, *Misho* style, *Kodo* style, and *Seizan* style. To enter into the peculiarities of these different styles would be almost like discussing some of the more trivial differences of religious sects. The main principles of the art are the same throughout. In the present work the *Enshiu* style has been chiefly followed, this being the most elaborate and the most popular of the more modern styles, but numerous examples have been taken from compositions of other Schools.

The *Enshiu* style was originated by a retainer of the Shogun Iyeyasu, called Kobori Totomi no Kami. He was a distinguished professor of the Tea Ceremonial, and became teacher of this accomplishment to the Shogun's heir, Iyemitsu. Compared with some of the other styles, that of Enshiu is characterized by a greater degree of artificiality or artistic affectation, and this makes it specially adapted for the purposes of a thorough explanation of the principles of the art.

PLATE VII.

EARLY STYLE OF FLOWER ARRANGEMENT (SHIN-NO-HANA).

PLATE VIII.

EARLY STYLE OF FLOWER ARRANGEMENT (SHIN-NO-HANA).

PLATE IX.

EARLY STYLE OF FLOWER ARRANGEMENT (SHIN-NO-HANA).

PLATE X.

LINEAL DISTRIBUTION.

AN analysis of Japanese flower arrangements shows that the lines or directions taken by the different stems or branches form the basis of all compositions. While European floral decorations are merely combinations of masses of colour, in which blossoms and leaves alone play a part, those of Japan are synthetic designs in line, in which every individual stem, flower, and leaf stands out distinctly silhouetted. The treatment followed may be likened somewhat to the methods employed of distributing carved foliage in architectural panels.

The surface of the water in which the flowers are placed is technically considered to be the soil from which the floral growth springs; and the designer must here convey the impression of stability and strength. However good the upper lines of the composition may be, a weak *springing* at the base deprives it of life and vigour; for it must be remembered that not flowers alone, but floral growth and vitality, are to be expressed in the designs. The direction of the stems at starting need not be strictly vertical, but, if curved, the curves employed should be strong ones, and all weak bends and angles should be avoided. As a composition consists of several main lines, there will be severall ines of *springing*, or origin. In some cases the *springing* lines are all united from the surface of the water to some distance above, after which they separate in tangential curves in different directions; in other cases, each stem-line is kept distinct, being separated from the others from the point of origin.

In the distribution of the principal lines of the composition from the point of their separation, the artist studiously avoids an equal-sided or symmetrical arrangement, but he obtains a balance of a more subtle kind, which is at the same time productive of a pleasing variety of form. Balance and harmony without repetition is a governing principle in this as well as in other Japanese arts. The lines of each stem, or, in cases where numerous slender stems are combined, the central lines of each group of stems, receive first attention. The triple arrangement, by which is meant that governed by three prevailing lines, may be taken as the original model for all arrangements (see Plate XI. A).

The three lines of such compositions may be called, with sufficient fidelity to the more quaint native nomenclature, *Principal*, *Secondary*, and *Tertiary*. The *Principal* line, as the name implies, is the central and longest line of the design, and this is made to form a double curve with the upper and lower extremities nearly vertical and in a continuous line, the general shape thus assumed being that of an archer's bow. The *Secondary* line should be about half, and the *Tertiary* line about one quarter, of the length of the *Principal*, supposing all to be straightened out; and these two lines are arranged on different sides of the *Principal* in graceful double curves of varied character. As a general rule, the *Secondary* line has a more vertical, and the *Tertiary* line a more lateral tendency; the former being on the outside of the arched bow formed by the *Principal*, and the latter making a counterpoise on its hollow side. According as the hollow of the *Principal* faces right or left, the arrangement is called a right or left composition. By changing the direction and giving a different character to the curves of these three lines a great variety of designs is produced, some of which are shown in Plate XI. B.

To produce a five-lined arrangement, two additional lines are introduced between the three previously named. The one placed between the *Principal* and the *Secondary* is called the *Support*, and the one between the *Principal* and *Tertiary* is called the *Sub-principal*. The *Support*, in length and importance, approaches more to the *Secondary* than to the *Principal*; while the *Sub-principal*, as its name would imply, in length and importance, approaches more the *Principal* than the *Tertiary*. In this way, it may be observed, a proper lineal balance and harmony is obtained; for supposing the three elementary lines to be valued 4, 2, and 1, according to their respective lengths and importance, it is plain that to preserve balance with the addition of two extra lines, that introduced between 4 and 1 should be longer and more approaching in value to 4 than that between 4 and 2, which should more approach the dimensions and character of 2. These additional lines, besides having different intermediary lengths, have other curves given to them, and are arranged so as best to fill the intervals with due regard to variety and balance. In the seven-lined arrangements two more extra members are added, one called the *Side-line*, and the other the *Trunk-line*. Their lengths are about intermediary, the *Side-line* being placed between the *Support* and the *Tertiary*, and the *Trunk-line* between the *Sub-principal* and the *Secondary*.

The different members of the above lineal arrangements have curious fancy

names bestowed on them by the different Schools. For the triple style such classifications as that of *Father, Mother,* and *Self;* or that of *Heaven, Earth,* and *Mankind,* are used; and for the five-lined style the terms *Centre, North, South, East, West;* or *Earth, Fire, Water, Metal, Wood;* and sometimes *Yellow, Red, Black, White,* and *Blue* are employed. The Enshiu School apply the following curious ideas to the different lines in an arrangement, namely: *The Heart, Help, The Guest, Skill, The Finishing touch;* referring to the different sentiments which should inspire the designer.

The general form of the above groups of three, five, and seven lines depends mainly upon the amount of curvature given to the *Principal* or centre-most line. In the simpler and less affected styles, this bow-like curvature of the *Principal* is slight and strong, but in the more exaggerated compositions, this governing line is boldly bent sideways from a point a few inches above the *springing*, and turned in a quick curve back again so as to bring the upper extremity vertically above the base, and to preserve the centre of gravity of the whole. The general form thus imparted is, as mentioned above, somewhat like that of a strung bow. From such a shape

the transition is not great to that produced in the arc of a bow by drawing it; a more violent lateral curvature with less vertical height being produced. This latter character is given to the *Principal* chiefly when used in positions where too much height would meet with actual or optical obstructions and be detrimental to a good effect. In floral arrangements placed below a shelf, or in front of a hanging picture which must not be hidden, such treatment is often followed. This modification in the character of the *Principal* necessitates corresponding changes in the directions and curvature of the auxiliary lines of the composition.

Up to the present, slight or violent as may be the bow-like curvature imparted to the *Principal*, it has been always presumed that its general direction—that is the direction of a line joining its base and top—is vertical. Thus, supposing a tri-lineal composition enclosed in a right-angled tri-angle, such tri-angle would be standing on its point, with the

Fig. 1.

hypotenuse, corresponding with the *Principal*, placed vertically (see Fig. 1). There is, however, another style of design applied to a large class of flower arrangements in which the *Principal* line of the composition has a horizontal or almost horizontal direction, and supposing a tri-lineal group of this style enclosed in a right-angled tri-angle the hypotenuse of the tri-angle would lie horizontally (see Fig.
2). This latter style is chiefly applied to flowers arranged in hanging
vessels, or in standing vases placed
on raised shelves, the intention of
such compositions being to suggest
floral growths on the edges of cliffs
or banks, which lean over laterally.
In such elevated arrangements it
often happens that one of the auxiliary lines on the side towards
which the *Principal* bends over, re-

FIG. 2.

ceives a decided droop and proportionate lengthening in order to emphasize the droop. Such drooping lines are technically called *Streamers*. The *Streamer* character may be imparted to either of the auxiliary lines, but redundancy must be avoided, and it is regarded as a fatal error to introduce *Streamers* on both sides of the *Principal*. The *Streamer* is very much used for floral arrangements in suspended vessels, in tall bamboo vases having side mouths, or in receptacles placed upon the edges of raised shelves. The prevailing idea in each of these cases is that the composition must suggest the wild growth on the edge of a bank or precipice over which the *Streamer* droops (see Plate XII. A).

The above description has been confined to three, five, and seven-lined designs. Single line and double line compositions, as well as those of nine, eleven, or more lines, are sometimes made, but their use is very rare. The double-line form is employed occasionally for the simplest of all flower arrangements, namely that of one flower and one leaf.

The different lines of a composition have hitherto been spoken of as if existing in one vertical plane parallel to the spectator, but actually, in addition to the directions mentioned to the right, left, upwards, and downwards, these lines have also directions of varying degrees forward or backward. In other words, the extreme

B

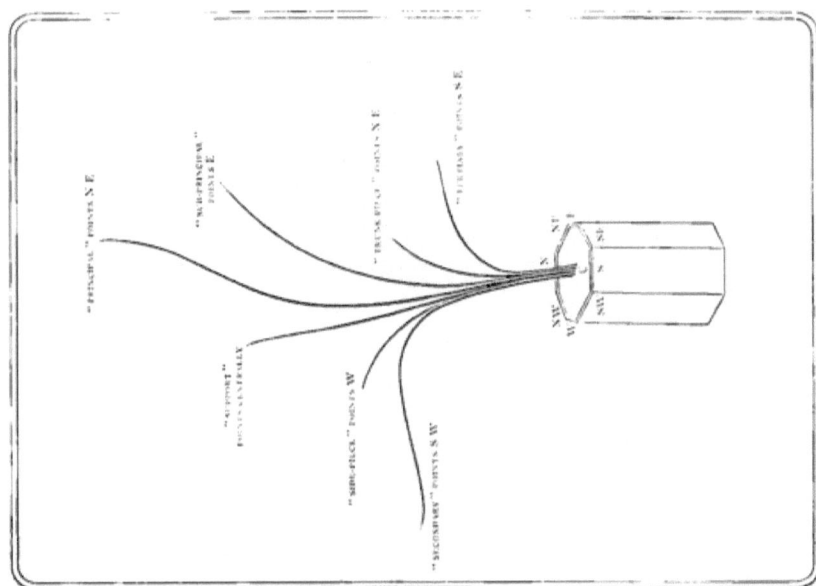

"PRINCIPAL" POINT N E

"SUB-PRINCIPAL" POINT E

"TERMINAL" POINT N E

"SECONDARY" POINT N E

"SUPPORT" POINT SECONDARY

"SEMI-PEAK" POINT S W

"SECONDARY" POINT S W

N E
N
N W
W
S W
S
S E

A

"SECONDARY" AS A "SERVICE"

"SUB-PRINCIPAL" AS A "SERVICE"

"TERMINAL" AS A "SERVICE"

"SUPPORT" AS A "SERVICE"

points of these lines would require a solid and not a plane figure for their enclosure. These directions are best explained by supposing a bundle of stems placed in a vase of octagonal plan and designating the eight different facets of the vase respectively as North, North East, East, South East, South, South West, West, North West: then, imagining the South face of the vessel immediately in front of the spectator, and bearing in mind that all the stems coincide for some little distance from their origin, the cardinal directions which they take after separation would be as follows (see Plate XII. B):—The *Principal* bends North East, the *Secondary* South East, the *Tertiary* South West, the *Sub-principal*, which is between the *Principal* and the *Tertiary*, turns East, and the *Supprt*, which lies between the *Principal* and *Secondary*, terminates centrally over the vase. Of the two extra lines used in seven-lined arrangements the *Trunk-line* bends North East, and the *Side-line* bends West. In this manner a pleasing variety back and front, as well as laterally, is given to the compositions, and they are not the flat arrangements which might at first sight be supposed from explanatory drawings. Though designed principally with a view to being seen from a point of view immediately in front, they obtain hereby solidity and variety, and their effect when regarded from points to the right or the left are also taken into consideration to some extent.

Certain errors in arranging the lines of floral designs are pointed out to be strictly avoided. The first is that of *Cross-cutting*, which is produced by allowing two or more lines visually to intersect one another, so as to form angles.

A similar fault, called *View-cutting*, exists when a stem or twig cuts across its parent stem or branch; but this is allowed for certain trees, as for example the Plum, in which such intercrossing is characteristic of the growth.

Another error, called *Parallelism*, occurs when two or more adjacent stems or branches are exactly parallel to one another, a fault much aggravated when the lines are besides of equal or nearly equal length.

Window-cutting, so named after the curved or cusped windows of the temples of the country, is the name given to an error occasioned by allowing stems to cross and recross in curves so as to form looped openings.

Lattice-cutting is an exaggerated form of *Cross-cutting*, and this error appears when several stems cross in parallel directions so as to suggest lattice work.

Another fatal error is that of the *Double streamer*, or drooped branch, used on both sides of the same composition. The above rules of prohibition, it will be observed, are for the most part similar to those followed in floral and other ornamental designs in the best styles of the architectural arts, being intended to prevent harsh, conflicting, or redundant effects and to be productive of harmony of line. They are illustrated in Plate XIII. A.

The various directions imparted to plants and branches of trees on the above principles are obtained, first, by a careful selection of suitable material; then, by twisting, bending, building together, and fixing at the base; and, lastly, by means of cutting and clipping off defective and superfluous parts. Some special methods of manipulation will be afterwards considered.

PLATE XIII.

B.

A.

SELECTION OF MATERIAL.

PON the general lines of composition already indicated, flower arrangements are made, sometimes with one species of tree or plant alone, and sometimes with a combination of two or more species. The use of many different kinds of flowers in one composition, though followed in the earlier styles, is opposed to the principles of the purer styles afterwards developed.

Combinations of two or three different species are, however, very common, and especially applied to vessels having two or three openings. In all compositions, single or combined, the special nature of the different materials employed is in each case carefully kept in mind, and anything at all suggestive of the inappropriate most scrupulously avoided. Important distinctions are made between trees and plants, and between land and water plants. The locality of production, whether mountain, moor, or river, considerably influences the arrangement adopted. Each flower has its proper month or season, and many flowers which are common to more than one season have special characteristics of growth, or of form, during the different seasons. These points of distinction are carefully studied, and are expressed in the artificial arrangements, within the limits of the art. Second flowerings, or flowers appearing in times other than their proper season, are discarded in flower compositions. As an example of this may be mentioned a late kind of Peach blossom, appearing in the Summer, which, beautiful as it is, may not be employed, because the Peach blossom is specially a flower of the Spring time.

In arranging two or more species in one composition, variety must be sought by combining trees and plants. In the case of three lines being used, the branches of a tree should never be "supported" on both sides by a *plant*, nor should a *plant* be placed in the centre with a *tree* arrangement on either side. This fault is called by a term which will be better understood if freely translated as *sandwiching*. In a triple arrangement it is plain that two branches of the same kind of growth must be used, but these must adjoin, and not *sandwich* the remaining one. As an example of a defective arrangement may be taken a composition with Irises (*plant*) in the centre, and branches of Azalea and Camellia (*trees*) on either side. A correct composition would be one with a Plum branch (*tree*) in the centre, with a Pine branch (*tree*) on one side, and a Bamboo stem (*plant*) on the other. For examples of the above correct and incorrect arrangements see Plate XIV. Some Schools allow occasionally the violation of this rule, and permit the supporting of a tree on either side by a plant, provided that the tree is a mountain tree and that one of the plants is a land and the other a valley or water plant.

The above rules were no doubt made principally to prevent the weak and insipid arrangements likely to be produced by the careless violation of the principles which they express, especially in the hands of the inexperienced. Plants, as before mentioned, are regarded as *female* with respect to trees, which are considered *male*, because the former are weaker and more graceful in character than the latter. A slender *plant* flanked on either side by *tree* branches would give a weakness of effect to the centre of the composition, and the reverse arrangement would give too much strength to the centre and weakness to the sides. In addition to this, such arrangements would have a more or less symmetrical character, and symmetry is disliked throughout the whole of the art under consideration. Like most arbitrary rules, such directions were often departed from by the more advanced professors, and there are even recognized exceptions which are universally admitted as correct. For example, two kinds of Pine branches may be used together with a Plum branch in a triple arrangement. Also in double compositions the Willow and Camellia are used together, and the *Celosia argenta* is often employed with the *Serissa fœtida*.

As previously stated, the branches and foliage of evergreens, and even of deciduous trees, are much used in floral compositions, the arrangement often being without a single blossom. It is, however, laid down as a general rule that no flower-bearing plant is to be employed with leaves only, nor must plants or trees which bear leaves at blossom time be used with flowers alone. The following are exceptions to this rule. The large leaved Chinese orchid has a flower, but it is very insignificant and grows below, and this plant is therefore technically treated as a flowerless one. The *Iris japonica* is sometimes arranged for its leaves only, before the flowers appear, and it then receives a special name. The leaves of the Summer *Narcissus*, sometimes

called by the fancy name of *Golden Pillar plant*, are faded and withered before the flowers appear, and they may therefore be removed and discarded.

All flower compositions must partake as much as possible of the spirit of the seasons in which they are used. Spring arrangements should be strong and powerful in line, like the growth of early vegetation; Summer arrangements must be full and spreading; whilst those of Autumn should be spare and lean; and those of Winter, withered and dreary.

Mention has been made above of certain fatal errors in combining the stem-lines of a flower composition, which are to be carefully avoided. There are other faults in arrangement which concern the distribution of leaves and flowers. Large blossoms should not be allowed to occur in regular steps or rows in a composition, and this error is called the *Stepped arrangement*. Another fault called the *Nail head style*, after the stiff metal flower ornaments used to cover nails in Japanese buildings, consists in a flower turned exactly full view to the spectator with leaves on either side. The term *Colour-cutting* is given to the error of placing blossoms of the same colour on either side of a single blossom of another colour; as, for example, a white Chrysanthemum flower between two red ones. This is only another example of the prejudice prevailing against *sandwiching* of any kind. The term *Dew dropping leaf* is given in a composition to a leaf which droops down in a feeble manner so that it could not support a drop of water; this is greatly objected to. The above errors are illustrated in Plate XIIIᴅ.

Three distinct characters are observed both in flowers and leaves. In flowers, there are the full blossoms, the half open blossoms, and the buds; and in leaves, the young green leaf, the full leaf, and the reddening or falling leaf. In flower arrangements with one material, as for example with Cherry or Peach blossoms alone, a different character of blossom is selected for the chief lines of the composition. For the *Principal*, full blown flowers will be used; for the *Secondary*, half open flowers; and for the *Tertiary*, buds are employed. Some designers, on the principle that the half open flower is more powerful than the full blown blossom, use the half open flowers for the *Principal* and the full blown blossoms for the *Secondary*. Straight leaves are considered strong, and curled or bent leaves weak; the strong flowers should be near the weak leaves, and the strong leaves near the buds or over blown flowers. A flower below a leaf is weaker than one above. In thinning out leaves in a composition, two strong leaves must remain for every weak one.

FLOWER VESSELS.

STANDING VASES.

AVING classified the different plants and trees which provide material for the Japanese arranger of flowers, it is necessary to describe the various sorts of flower vessels which are employed for receiving the compositions. The form and character of such vessels considerably influence the nature of the floral arrangements placed in them. These receptacles are of a diversity of shapes and of different materials, such as wood, porcelain, pottery, bronze, brass, iron, and basket-work. Without attempting to trace the development of the different art-industries as applied to vessels for receiving flowers, it will be necessary to refer to the various shapes, proportions, material, and decoration of such vessels, so far as they control or are governed by the floral arrangement adopted. The most ancient receptacle used was a long-necked earthenware or bronze vase of considerable height, and the flower composition placed therein was made proportionately tall. The difficulty of balancing such high arrangements led to the use of shorter and broader vessels and to a corresponding lowering of the compositions. This fact shows that from the earliest times a recognized connection of proportion existed between the floral group and the vessel in which it was placed.

Fig. 3.

It is important to note that the surface of the water in which the flowers are placed is technically regarded as the soil from which the artificial arrangement springs, and the designer must here convey the impression of a stable origin. With this principle in view, and guided also by the necessity, previously mentioned, of avoiding too great a height in flower compositions, a sort of broad-mouthed vase became the favourite form used. Of this kind there are many shapes, generally in bronze, and varying from a low saucer-like vessel to a trumpet-shaped form, supported sometimes on short legs, and sometimes upon castings, representing such subjects as rocks, water, spray, or an animal group.

Other kinds of vessels corresponding more to the ordinary vases employed in the West, having necks and oval or cylindrical bodies, are also much used. Among these, the vases with tall narrow necks are employed for the simplest arrangements such as are displayed in the tiny Tea Rooms where rules of austerity necessitate very light and quiet floral compositions. Vases intended for elaborate flower arrangements are always necessarily of the wide-mouthed kind, to ensure an appearance of stability at the base of the flower stems. A few of these different vases are shown in Plate XV., but there exist a number of other shapes in common use, besides several fancy forms which are occasionally employed. With the ordinary tall vase, whether of wide or narrow mouth, the height of the flowers is generally fixed as one and a half times that of the vase. Some Schools increase this proportion, giving double the height of the vase for the floral composition.

VESSELS FOR WATER PLANTS.

For the purpose of displaying, in a suitable and suggestive manner, water plants and grasses, low tub-like vessels came into use from quite early times. There are two principal kinds employed, called respectively the *Sand-bowl* and the *Horse-tub.* The *Sand-bowl* is a broad shallow vessel, oblong, poly-gonal, or sometimes oval in shape, made of bronze or porcelain, and containing a layer of pebbles or sand covered with water. The *Horse-tub* is, as its name implies, a shallow tub of wood, circular in plan, and generally lacquered black. Its use as a flower vessel is said to have originated during the exigencies of war time, when a famous general of æsthetic tastes, in the leisure of a long campaign, first used a common horse-tub for arranging

FIG. 4.

flowers in. Whereas it is a rule with other kinds of flower vases that the water shall not be visible, the water in vessels of the *Sand-bowl* and *Horse-tub* kind is intended to form part of the composition. The idea to be conveyed is that of a

broad surface of water or marshy ground, and the plants and trees used are only
those which are associated with the propinquity of water. Water plants are chiefly
employed, but sometimes plants or trees which grow on the banks of streams are
also allowed. The use of such water vessels
is moreover chiefly confined to the warmer
months of the year, when the sight of a
broad surface of water is grateful and refresh-
ing to the eye. A different rule of proportion
between the flower composition and the vessel
is followed when these broad shallow receptacles are employed. The height of the floral
composition is made about one and a half
times the breadth of the vessel, and is not
regulated by its height as with the ordinary
standing vases.

Fig. 5.

It is customary with the above-named
water-bowls to use fancy fasteners of metal,
to hold, or give the appearance of holding,
the base of the flower stems. These will be
described afterwards when the whole question
of fasteners is discussed. Examples of *Sand-bowls* and *Horse-tubs* are shown in Plate
XVIA. A special water vessel requiring mention is that which goes by the name of
the *Long-boat*, not from any particular resemblance in form to a boat, but from the
fact that it is very long in proportion to its width, and is made of white wood. In
shape, the *Long-boat* resembles a *Sand-bowl*, and it is used in a similar manner,
with sand, pebbles, rocks, and water.

Another form of standing vessel sometimes used for plants is the *Well-frame*,
a sort of box-shaped vessel of worm-eaten wood, with a capping piece crossing at
the angles like an Oxford-frame, the whole resembling the wooden enclosure round a
Japanese well (see Fig. 5).

The rustic character of this kind of vessel requires that the flowers placed
therein should be of the simplest kind. In the example illustrated, a plant called
Togiri (*Clerodendron squamatum*) is combined with a long slender grass called **Susuki**
(*Eularia japonica*).

FLOWER BASKETS.

The famous Regent Yoshimasa is said to have been the first to employ the plaited basket as a receptacle for flowers. These baskets, made of reeds, stems of creeping plants, cane, or bamboo, were of Chinese origin, and much prized in Japan,

where high prices were given for antique specimens. To the present day the more mellowed with age these vessels appear, the more they are valued. A Chinese artizan, named Hokoji, is said to have introduced their manufacture into this country. He presented one of his own make to the retainers of Yoshimasa, with a humble request that it should be embellished with an ornamental stand when placed before the Regent. Yoshimasa is said to have been so pleased with its simple elegance that he ordered it to be put at once on the dais without any stand or tray. From this arose the custom of dispensing with a tray under *Flower-baskets*, though used under other flower vessels. A special kind of *Flower-basket*, with a large oval handle arching over the top, is still called the *Hokoji* shape, from the name of its first maker; and another kind, with a straight neck and no handle, is called the Reishojo shape from the name of his daughter, Reishojo. The Hokoji-shape basket is shown in Fig. 6,

FIG. 6.

arranged with a Convolvulus twined round the handle. Flowers placed in such baskets are fixed in bamboo tubes containing the water, which are hidden within.

Another popular story, in connection with the use of baskets as flower vessels, is that the famous philosopher Rikiu started the fashion on the occasion of a visit to the cherry blossoms on the banks of the river Katsura, near Kioto, when he made use of a common trout-basket to arrange gathered flowers in. There is a special basket still in use called either the *Katsura-basket* or the *Trout-basket*. Besides the orthodox shapes alluded to, there are various fancy forms of baskets employed both for standing and hanging. Of these the following principal examples may be mentioned:—

The *Half-plaited-basket*, a cylindrical basket in which the vertical threads project two or three inches above the horizontal plaiting, as if unfinished in making.

The *Rain-coat-basket*, a hemispherical basket with a frayed top hanging over the edge of the mouth, and somewhat resembling the straw collar of a Japanese farmer's rain-coat.

The *Wool-basket*, a rudely shaped cylindrical basket with a long handle, such as is used in collecting the wool of the cotton plant.

The *Bait-basket*, a small plain basket with a handle, similar to that used for carrying fishing-bait.

The *Horned-basket*, a basket of tall cylindrical form, with two horn-like projections on the top, as handles.

The *Square-basket*, a basket of simple cubical shape with no handle.

The *Long-basket*, a tall, thin, tube-like basket.

The *Sosen-basket*, a basket of pyramidal shape, having four sloping sides gathered to a circle at the mouth, and with a high arched handle. The name Sosen is that of its inventor. It is shown arranged with Chrysanthemums in Fig. 7.

Among special baskets for hanging or hooking, may be mentioned the following :—

The *Horn-shaped-basket*, a basket shaped like a bull's horn.

Fig. 7.

The *Cicada-shaped-basket*, so called from its resemblance to the body of an insect.

The *Butterfly-shaped-basket*, a basket which has a short cylindrical neck, and a broad arched body curving out to points at the bottom, the whole form being suggestive of a butterfly with open wings.

The *Sock-shaped-basket*, a basket of an irregular oval shape below with a top opening towards one end, somewhat resembling in shape a baby's sock. It is illustrated in Fig. 8, arranged with Willow and Narcissus.

The *Gourd-shaped-basket*, a basket woven in the form of a gourd.

The *Daikoku's-bag-basket*, a basket resembling in shape the bag carried by the Japanese God of Riches, *Daikoku*.

The *Arima-basket*, a basket of cylindrical form with a side mouth, similar to the hanging bamboo basket. The name is derived from a village called Arima.

For certain suspended arrangements of creeping plants a deep net-work basket, something like an English waste-paper basket in shape, is often used. Within this a low flower vase is placed from which the creeper rises, being arranged so that part is seen through the netting of the basket, and part hangs over the side in a *streamer* or trail. In all cases

Fig. 8.

flower baskets require an inner vessel for holding the water in which the flower stems are placed, and this generally consists of a concealed tube of bamboo. Examples of various flower baskets are to be seen in Plate XVIB. and Plate XVII.

BAMBOO VASES.

To Yoshimasa's patronage is also attributed the original use of flower vases made of bamboo tubes. As first introduced, these were simple cylinders of thick bamboo, cut near the root, about a foot or more in height, and four or five inches in diameter, the bottom being closed by a natural division. The facility with which such vessels could be cut into different shapes led to the invention of a variety of forms, each bearing a fancy name and specially adapted to different styles of flower arrangement. Portions of the sides were notched out, and side apertures were introduced, often in several stages, so as to allow of two, three, or five compositions in one vase. The different kinds employed are so numerous as to require tabulation, and many of the names almost defy translation; they are therefore given in Japanese. The meanings of these names are explained as far as possible, and it will be seen that they refer chiefly to some fancied resemblance in general shape, or in the form or number of the apertures, to other native objects. They are as follows :—

FIG. 9.

Shishiguchi-gata.—Lion's-mouth-shape; a cylinder from ten to fifteen inches in height with a square side-mouth about 3 inches deep.

Noborijishi-gata.—Rampant-lion-shape; a higher cylinder with side mouth as above.

Tabimakura-gata.—Travelling-pillow-shape; a very short vase with small side slit.

Utaiguchi-gata.—Singing-mouth-shape; so named from a splayed form given to the side opening suggesting the incline of the lips in singing.

Waniguchi-gata.—Shark's-mouth-shape; so called from teeth-like prominences given to the bottom of the side aperture.

Fukurokuju-gata.—Named after a Japanese magician with a very long cranium, on account of the low position of the side mouth.

Karamon-gata.—Chinese-gateway-shape; a high vase with one of the side openings rounded, resembling a Chinese arch.

Gammon-gata.—Wild-geese's-gateway-shape; a vase with curved side-opening somewhat like the above, the name being taken from that of a famous arched sea rock called the *Wild-geese's-gateway.*

Kawataro-gata.—Named after a fabulous animal called *Kawataro* or *Kappa* something like a monkey, said to inhabit lakes.

Seiro-gata.—Distilling-vessel-shape; a high vase with two small side openings one above the other.

Kawara-gata.—Tile-shape; so called from its resemblance to a half cylindrical capping tile.

Ro-gata.—Oar-blade-shape; resembling the blade of a Japanese oar.

Toro-gata.—Lantern-shape; supposed to resemble a stone standard-lantern.

Anko-gata.—Ray-fish-shape.

Daruma-gata.—Seated-hermit-shape; this vase spreads out at the bottom and has an arched side-opening near the top, suggesting the form of a hermit's cowl.

Noborizaru-gata.—Climbing-monkey-shape; this is a high vase with a very long deep cut in the middle leaving two short cylinders at top and bottom.

Daibutsu-gata.—Seated-Buddha-shape; so called from its supposed likeness to a seated Buddhist statue.

Enko-gata.—Monkey-shape.

Katoguchi-gata.—Cusped-opening-shape; so named because of the cusped form of the side-opening.

Toro-gata.—Mantis-shape.

Gojiu-gata.—Five-storey-shape; a very high vase with five side openings.

Hioki-gata.—Icicle-shape; so called from the leg-like cuttings at the bottom of the vase, in the shape of icicles.

Mitsunashi-gata.—Three-legged-shape; in this vase the bottom of the cylinder is cut away leaving three legs remaining.

Torikago-gata.—Bird-cage-shape.

Teoke-gata.—Hand-pail-shape; resembling a Japanese hand pail, two deep apertures being cut exactly on opposite sides so as to leave a handle-like strip remaining.

Tegine-gata.—Pestle-shape.

Usu-gata.—Mortar-shape.

Shakuhachi-gata.—Flute-shape; a long thin tube of bamboo, slightly bent like a native flute.

Hashigui-gata.—Bridge-post-shape; supposed to resemble the newel of a wooden bridge rail, having a deep square slit in the middle.

Miotsukushi-gata.—Beacon-light-shape.

Nijiu-giri-gata.—Two-storey-shape; a vase with two side openings one above the other, in addition to the top opening.

Sanjiu-giri-gata.—Three-storey-shape.

Tsurube-gata.—Bucket-shape; named after its supposed resemblance to a well-bucket.

Tsuru-kubi-gata.—Crane's-neck-shape; so called from the length and depth of the side cutting which leaves a long thin neck of bamboo suggestive of a crane's neck.

Tsurigane-gata.—Bell-shape.

Koma-gata.—Spinning-top-shape.

Tarai-gata.—Tub-shape.

Horagai-gata.—Conch-shell-shape.

Taki-nobori-rio-gata.—Cascade-ascending-dragon-shape; a high bamboo vase cut into a long spiral, supposed to resemble a writhing dragon.

Eboshi-gata.—Ceremonial-cap-shape; named after its resemblance to a Japanese Court-cap.

Jikiro-gata.—Food-box-shape.

Nijiu-yagura-gata.—Two-storey-castle-turret-shape; so called from square embrasure-like side openings near the top.

Hatomune-gata.—Pigeon-breasted-shape; so called from a bend in the bamboo cylinder giving it a pigeon-breasted appearance.

Rikkwan-gata.—Pan-pipes-shape; a row of small bamboo tubes of different heights tied together with cord and fixed on a stand.

The invention of most of these bamboo vases is attributed to different professors of the Tea Ceremonial. As will be perceived in the above list, assisted by the illustrations, the breadth, depth, and roundness or squareness of the side apertures, as well as their number, and the total height of the vase, suggest the names for the different vases. Many of them are provided with a circular nail hole on one side, near the top, for hanging purposes, and such vessels can be used at option, either hooked to a nail, or standing upon the floor of the alcove. The tall kinds having open tops are invariably used standing.

Another variety of bamboo vase not previously mentioned consists of three or more bamboo cylinders of different heights attached in a line, and named *The Row-of-Piles*, after their resemblance to a row of pile heads. Many of the above mentioned bamboo vases are illustrated in Plate XVIII.

There also exists what is called the *Verdant Bamboo Vase*, being a vase of one of the above shapes, freshly cut from a growing bamboo stem, with twigs of green leaves remaining on it. In such a vase the intermediary knots or divisions are left intact, and small apertures are introduced in the side for filling in water and other preservatives against speedy withering.

Japanese flower vessels may be broadly divided into three kinds, those used for standing upon a dais, table, or shelf; those intended for hooking against the wall or against a pillar; and those suspended by chains or cords from a ceiling or beam.

The vessels hitherto described belong to the *standing* kind, with the exception of some of the baskets and certain of the bamboo vases just enumerated, which can be used either for standing or hanging. In order to distinguish clearly between flower vases which are hooked to a pillar or wall surface and those which are hung by chains or cords, the former will be called in future *Hooked-vessels,* and the latter *Suspended-vessels.*

HOOKED VESSELS.

Hooked-vessels are of various kinds, from the chrysalis-shaped root of a bamboo to the form of a shell, gourd, or melon. They are invariably short, compared with the standing vases, and with a few exceptions the absence of flatness below

suggests their method of use. Among the bamboo vases, those of little height and with narrow side apertures, such as the *Lion's-mouth-shape* and *Travelling-pillow-shape,* are used mostly as hooked vases. A lateral direction is given to floral compositions placed in hooked vessels, the idea suggested being that of flowers hanging over a cliff. For Tea Rooms, where a severe and rustic style of flower composition is preferred, curious shapes of vases are pressed into use, among which may be mentioned those of the *Octopus-pot,* a coarse irregular shaped earthenware jar used by fishermen for holding the octopus; the *Iron-pot,* a rough iron pot-shaped vessel somewhat like a martin's nest; and the *Decayed-stump,* a piece of decayed wood hollowed out as a vase. The *Gourd* is also a favourite form for hooked vases, the mouth being sometimes cut in the side and sometimes at the top. It is said to have been first used for flower arrangements by the philosopher Rikiu, who once extemporized a floral design in a wine-gourd which he took from an itinerant priest at the temple of Sumiyoshi in Osaka.

Fig. 10.

As previously mentioned, many of the woven baskets employed as flower vessels belong to the hooked class. These are to be found described under the head of flower-baskets, and are illustrated in Plate XIX.

As a background to the *Hooked-vessels* and originally intended to protect the pillar or papered wall-surface from staining or abrasion, narrow oblong tablets of wood are often used. They are sometimes made ornamental, being lacquered and inscribed with verses in gold letters. Some are plain oblong tablets about four inches broad and three or four feet long; others are wedge-shaped, tapering towards the top; and others have curved sides. They are provided with a long narrow slit down the middle for sliding to different heights over the iron nail or peg by which they are held to the wall or pillar and to which the flower vase is hooked. In some cases these tablets are hinged in the middle to allow of folding up when out of use. They are often made of segments of bamboo flattened out and polished or lacquered. Examples of these hanging tablets may be seen in Plate XXA.

SUSPENDED VESSELS.

Suspended vessels are those hung by a cord or chain to the ceiling or lintel of a recess. Belonging to this class is a crescent-shaped vase of pottery or bronze called the *Crescent-moon*. The horns of the crescent are made almost to meet and are suspended from above by a connecting ring and single chain. The other kinds being of more elongated form are hung from both ends by double chains or cords. The simplest of these are bamboo tubes splayed off at the ends, hollowed out in the middle, and hung horizontally so as to suggest the form of a boat or punt; others are of bronze, shaped in exact resemblance to a ship or junk. Yoshimasa is said to have conceived the idea of boat-shaped vases whilst observing children sailing toy boats filled with flowers. Another story attributes the first use of such vessels to the famous philosopher Soami who on a hot summer day, to please his patron Yoshimasa, took a bronze vessel of accidental resemblance to a boat, and by his manner of arranging the stems of the flowers therein, conveyed the idea of a sailing vessel. The Regent was so pleased with this new departure in flower arrangements, that Soami devoted his attention to the matter, and drew up certain rules with special regard to boat arrangements. The different terms then introduced refer both to the form or decoration of the boat employed, and to the direction or character of the complete floral composition. First, out of compliment to Yoshimasa, who was then in retirement at Higashiyama, the names of his ten pleasure boats were applied to different forms of flower-boats. These were as follow;— *Dragon-head-boat*, *Dragon-boat*, *Phœnix-head-boat*, *Phœnix-boat*, *Lotus-flower-boat*, *High-wave-boat*, *Cloud-boat*, *Chinese-ornament-boat*, *Divers'-boat*, and *Seaweed-gathering-boat*. These names refer, except in the last two cases, to the ornament carved on the prow or sides of these Royal pleasure boats, both for decorative purposes and intended as a charm against accident. In the *Flower-boats* used to represent them, such designs were either made on card-board fixed in the end of the vessel, or were actually cast upon the side of the bronze vessel.

The simpler boat-vases of bamboo seem to have been of later origin than the bronze boats previously described. They have various distinguishing names according to their length and general shape. Such are; the *Boat-shaped-boat*, a very short vessel; the *Ocean-boat*, a long vessel with a high stern; the *Roofed-boat*, a vessel in which the form of the cutting suggests a roof over the deck; the *Oil-funnel-boat*, a long narrow boat; the *Companion-boats*, two boats connected side by side, one a little ahead of the other; and

FIG. 11.

the *Canal-boat*, a punt or barge-shaped vase. A somewhat similar kind of vessel consists of several small bamboo tubes of irregular lengths tied together to resemble a *Raft*, and hung horizontally. The centremost of these tubes is sometimes furnished with an open-ing to receive the flower stems, but in other cases the whole arrangement is merely a suspended stand, upon which is placed a small bowl or vase holding the flowers. For the latter purpose *Rafts* are sometimes made of squared bars of wood bound together, instead of bamboo tubes. An illustration may be seen in Fig. 11, in which the Raft supports a basket of cherry blossoms supposed to be suggestive of river scenery at Arashiyama where flower-laden boats float down the stream in the cherry season.

Various other fancy shapes of suspended boat-like vessels are used. Of these may be mentioned;—the *Koto-shaped boat*, a broad and slightly arched vessel with square ends, resembling an inverted Japanese *koto* or harp; the *Snow-shoe-boat*, a vessel in the shape of a large snow-shoe with a little frame upon it within which is placed a small flower vase; the *Decayed-wood-boat*, a boat-like vessel formed out of a hollow-ed piece of old timber, the *Chinese-brass-boat*, a fancy vessel of brass, with a silver chain, and an anchor of black metal hanging to the side; the *Basket-work-boat*, a boat of metal basket-work. Also a curved cane-work tray, oblong in shape, hung from the ends, and carrying in the middle a little bowl of flowers, is sometimes used.

An important theory in boat arrangements is, that they ought always to be suspended in an elevated position, both with a view to preserve the idea of a floating vessel, and also to prevent the possibility of seeing the water which they contain.   It is held to be a great violation of taste to allow the water which is necessary for preserving the plants in a flower-boat to be seen, because water visible within a ship would be suggestive of a leaking or wrecked vessel, and would be consequently considered ominous.   Sometimes the idea of a stranded or beached boat is purposely conveyed by a flower vessel which is placed upon the dais instead of being suspended. In this case the vessel should be raised upon a stand of some kind so as to place its upper surface above the eye level of seated visitors.   Such standing boat-vases are supported upon two wooden rollers or upon a light frame of cross-pieces.   Before leaving the subject of boat-shapes, allusion must be made to a standing vessel called the *Long-boat* which is sometimes employed for elaborate arrangements of plants and grasses.   This vessel appears to be called a boat simply on account of its narrow length and the fact that it is made of plain white wood.   It resembles a *Sand-bowl*, being an oblong tray-like vessel with short legs.   It is five feet long and about one foot wide and is only used for very large recesses on special occasions.   The boat-vases described above are illustrated in Plates XXI and XXII.

The classification given refers only to the different shapes and sizes of the boat-like vases.   Other terms are used to designate the manner of hanging the vessels and of arranging the flowers within, so as to convey different nautical ideas.   The three principal arrangements are those of the *Outward-bound-ship*, the *Homeward-bound-ship* and the *Ship-in-port*.   Besides these there are other designs known as the *Distant-ship*, the *Swiftly-sailing-ship*, the *Becalmed-ship*, and the *Branch-laden-ship*.   These different fancies are conveyed first by the direction, right or left, and backward or forward, given to the prow of the vessel; and secondly, by the distribution of the different lines of the flower composition.   Even the length of the suspending chain and the distance or proximity of the arrangement from the observer is governed by the style adopted.   According to the principles of lineal distribution which apply to all suspended flower designs, the *Streamer* holds an important place in the above examples.   It is in such cases intended to suggest the long bent oar which in Japanese boats trails back towards the stern.   This floral line must not be too powerful, as it represents the idea of an oar dragging in the water.   The central flower stem stands for the single mast of a junk with or without sails, and the subsidiary stems indicate the other sails and rigging of the vessel.   This will be better understood by describing one or two of the arrangements in detail.

Homeward-bound-ship.   In arranging a boat of flowers in this form the prow of the vessel is turned towards the left, which, in superior rooms is the host's side of the chamber, in order to convey the idea of home-coming. The central stem of the floral arrangement is high and full, curving towards the helm, so as to indicate a ship in full sail, and a *Streamer* hangs over the front side sloping back towards the stern on the right. The above is a favourite device on occasions of rejoicing for a safe return, or when a son or daughter-in-law is being received into the family. Some say that this method of arrangement should only be employed from noon to dusk.

Outward-bound-ship. This is an arrangement exactly the reverse of the former, the vessel having its prow turned to the guest's side of the chamber, on the right. It is adopted at parting gatherings in token of wishing good-speed to those setting out on a journey. It is said that this style of composition should be employed only from morning to noon.

Ship-in-port. In this arrangement the vase has the same direction as the *Homeward-bound-ship*, but the floral design is kept small and straight, so as not to suggest wind or motion, and the *Streamer* hangs over the further side of the vessel. Such a disposition of the flowers should not be effected excepting during the hours of evening.

Swiftly-sailing ship. The direction of the vessel in this design is to the right, or outwards; the floral arrangement is full and bent, but no *Streamer* is used.

Branch-laden-ship. The direction of this vessel is inwards, or towards the left, and the floral arrangement is kept short and close, and consists of small flowers, such as daisies and carnations, which are not allowed to project beyond the limits of the vase itself. The idea suggested is that of a ship loaded with timber or tree branches.

These different styles of composition are shown in a skeleton form in Plate XXIII. Other special rules for hanging boat-vases will be considered afterwards, when the general question of the position of flower arrangements in a chamber is discussed.

Suspended vessels called *Well-buckets* are often used in pairs hung over a pulley by a thick silk cord. One of the buckets is allowed to rest on the floor, or in some cases upon a frame designed in imitation of the railing or boxing round a well, and the other is suspended in the air.

To Rikiu is attributed the first use of such flower vessels, the idea coming to him whilst he was observing a convolvulus twining round the bucket of an old well. A similar pair of buckets are occasionally employed without the pulley and suspending rope, one being placed balanced on the edge of the other so as to leave only a portion of the lower one uncovered for the insertion of flowers. In this case the rope is arranged in a flat coil as a stand for the lower vessel. Buckets used in this way are always flat-sided to ensure stability, but for the ordinary suspended arrangements cylindrical as well as square buckets are employed. These vessels are of plain wood, of wood lacquered black, or of worm-eaten or decayed timber. For the most handsome kind in black lacquer, a chain of silver or a red silken cord should be used; to those in plain wood a plainer cord may be attached; and in the case of the rustic buckets, of decayed wood, a common hemp rope or even an iron chain may be substituted. Single buckets are occasionally to be seen standing upon a low table or decayed slab of wood, or hung by a single bamboo rod. (see Plate XXIa.).

Porcelain buckets and pulleys, although not uncommon, are of quite modern introduction, and not according to rule.

Other fancy vessels suspended by cords or chains are sometimes employed. Among the bamboo vases in Plate XIXa. will be seen one which is suspended by a chain, like a lantern. Suspended baskets, distinct from the hooked baskets previously described, are not uncommon. Another example is the suspended net-work basket previously described on page 60.

The inverted bronze bell suspended by a chain is another vessel occasionally used for holding arrangements of Wistaria flowers. (see Fig. 14).

A curious form of suspended flower vase is the inverted umbrella, an exact imitation in bronze of a Japanese umbrella. (see Fig. 12).

A large sea-shell hung by a single cord forms a favourite receptacle for very simple flower arrangements.

FIG. 12.

FLOWER CHARIOTS.

Belonging strictly speaking to the class of standing vessels, but sufficiently striking and important to require special notice, is the *Flower-chariot*, which figures so often in pictures on painted screens and other decorative objects. At certain festivals and processions it appears that large tubs full of richly arranged flowers were drawn upon wheeled chariots handsomely ornamented. The idea was adopted for flower arrangements placed in very large recesses, where great size and display were required. The length of the *Flower-chariot* is four feet six inches from the back to the end of the shafts; the wheels are about eighteen inches in diameter, and the flower tub which the chariot carries, is about sixteen inches high. Both vehicle and flower tub are lacquered black and furnished with silver fittings. The flower compositions are made very full and high. The *Flower-chariot* is illustrated in Fig. 13.

Fig. 13.

A somewhat similar flower receptacle called the *Water-carrying-cart* is also employed. This vehicle is a flat truck with four low wheels, and no shafts, like a child's toy-cart, bearing a small bucket in which the flowers are arranged.

FLOWER FASTENERS.

The subject of *Fasteners* for floral arrangements is one belonging to the *Technique* of the art under consideration. To a great extent, however, the methods of fastening are treated as a part of the decoration of the compositions, and as such they are closely connected with the different flower vessels employed, and require notice in the present context.

As before mentioned, the *Springing*, or point of origin of the floral group, is of great importance, and the firm and skilful fixing of the stems or branches in the vessel which holds them is one of the most difficult parts of the manipulation. Ordinarily, the stems are held in position by small cylindrical pieces of wood fitting tightly across the neck of the flower vase, and having a slit, wider above than below, for threading them through. The wedge-shaped form, wider towards the top, which is given to the slit, allows slightly different inclinations to be imparted to the several branches. The fastener should be fixed about half an inch below the surface of the water, the level of which is made to vary according to the season, and it should not be visible from the front of the vessel. If the vase used be a lacquered one, paper should be placed between its surface and the ends of the fastener, to prevent scratching. In some large mouthed vessels, and in the *Flower-baskets*, the flower stems are fixed in concealed tubes of bamboo which hold the water and the fasteners. Some Schools affect a rustic simplicity in their appliances and employ a naturally forked twig to hold the flowers in position.

For arrangements of water plants in neckless vessels such as *Sand-bowls* or shallow *Tubs*, other sorts of *Fasteners* are necessary, which are hidden below the sand or pebbles which such vessels contain. One kind consists of a sheet of copper perforated with holes of different sizes to receive the extremities of the different stems. Another *Fastener* is made of rings or sections of bamboo of varying diameters attached to a wooden board, the stems finding lodgment in the sockets thus formed, and being further held in position by the pebbles which cover them. Occasionally a *Fastener* called the *Whirlpool*, and consisting of a spiral hoop of metal placed vertically, is employed.

For arrangements in these shallow vessels there are a number of fancy *Fasteners* in common use which are in many cases merely ornamental, the hidden contrivances just described, buried below the sand or pebbles, doing the real duty of holding the stems in position. The principal of these ornamental fasteners are as follow :—

The *Horse's-bit-fastener* is an exact counterpart of a ponderous Japanese bit. Its use originated with the employment of the *Horse-tub* as a flower vessel, and to this kind of vessel it is chiefly confined. The linked character of this fastener allows of its being folded in a variety of ways, so as to leave loops of different size to encircle the flower stems. With the scrupulous minuteness of detail which characterizes the art under discussion, the floral designer has classified these different ways of using the bit, giving names to each bar, plate, and loop of iron, and inventing terms for the various methods of folding. The principal arrangements are shewn in Plate XXIVᴀ. The use of the *Horse's-bit-fastener* is prohibited, however, for floral designs placed in the ornamental recess of a chamber of superior class; and if it be introduced into a flower arrangement in such important rooms, the composition must not occupy the principal position.

The *Crab-fastener* consists of a metal crab or pair of crabs. If one crab be used, it should be disposed so as to contrast in character with the flower arrangement, by which is meant, that if the composition be high and powerful, the crab must be placed in a low and unobtrusive position, but if the flower design be broad and wanting in vertical strength, the crab must be raised in a climbing attitude. If a pair of crabs be used, one must be elevated and the other lowered in position, or, to adopt the quaint phraseology of the floral art, one must be *male* and the other *female*. As the representation is that of a land and not a sea crab, this kind of fastener may be used with land as well as with water plants.

The *Hare-fastener* is a bronze hare in miniature. It may not be affixed to water plants, and is specially suitable for arrangements of wild plants and grasses such as the Lespedeza, Rush, and Elecampane.

The *Pair-of-carp-fastener* consists of a pair of metal fish designed in the position of two carp sporting together. This fastener is, as might be supposed, only used for water plants.

The *Dragon-fastener* is a metal ornament in the form of a writhing dragon, and, as the dragon is a mythical monster belonging to all elements, the use of this fastener is not limited to any particular kind of plant.

The *Tortoise-fastener* consists of one or two tortoises in bronze arranged in different positions.

The *Water-fowl-fastener* is generally a metal imitation of a pair of Mandarin ducks. It is occasionally attached to water plants.

The *Frog-fastener* needs no special explanation, except that, representing an amphibious animal, it may be used with both land and water plants.

The *Anchor-fastener* is specially intended for use in suspended boat-shaped vessels. It is incorrect to fix it in a vase representing a stationary ship, as in such a case the anchor would not be visible.

The *Knife-fastener* is a metal knife or dirk such as is worn in the wooden sheath of a Japanese sword, and owes its original use as a flower fastener to a floral arrangement once hastily extemporized by a famous artist named Oribe, in which, having no other fastener at hand, he used his knife for the purpose. The *Scissors-fastener;* the *Pipe-fastener,*—a long metal tube with a small bowl; the *Weight-fastener,* —an oblong metal paperweight; the *Chain-fastener,*—a short chain disposed in a bunch; and the *Kettle-stand-fastener,*—a small iron ring and tripod used for supporting the kettle over the charcoal brazier;—these also are all occasionally employed. The principal of the above fasteners are illustrated in Plates XXIV and XXV.

A special kind of fastener called by the Japanese *Jakago* needs separate notice. The native name *Jakago* refers to the long sausage-shaped bags of bamboo basket-work which are filled with bowlders and laid in fascines at the sides of rivers to break the current and protect the banks. They are a common feature in river scenery and have therefore come to be imitated in flower arrangements intended to be suggestive of the presence of water. The *Jakago-fasteners* are long cylindrical baskets with closed and rounded ends which are laid in shallow basins together with ornamental stones or rocks, and besides being decorative they serve to hold the stems of the plants artificially arranged within. An illustration of their use may be seen in Fig. 26, where two are shown combined with the *Kerria japonica,* the whole being intended to represent a view of the river Tama near which these flowers abound.

CHOICE OF FLOWER VESSELS.

The Japanese flower artist recognizes a distinct and important connection between the floral composition and the receptacle in which it is arranged. Some points of relationship, such as that of proportion in height or breadth between the two, and the distinction between vessels used for land plants and those suitable for water plants, have been already mentioned. But there are other more subtle harmonies in taste and sentiment which are carefully observed. A flower vessel, being in itself a work of art, may possess different aesthetic characteristics, such, for example, as rustic simplicity, elegance, or richness. Its shape, material, or decoration may also convey to the imagination various feelings and mental associations. Flowers, too, possess different qualities in form, colour, or growth, and are in all countries associated with special poetical attributes. Particularly is this the case in Japan where hardly a fête or pastime exists in which flowers do not play a part, and where almost every blossoming tree has some romantic tradition of its own. Apart from the character intrinsically belonging to the flowers themselves, their artificial treatment in the flower art under consideration imparts to them other characteristics varying with the style of arrangement adopted. Thus one composition may be full and luxurious, whilst another design made with similar flowers may be simple and even austere.

Considering, then, that both flower arrangements and flower vessels are works of art capable of distinct artistic expression, it becomes a matter of importance that the spirit of the one should accord with that of the other. This harmony of sentiment need not necessarily be one of complete unison; on the contrary, it is often produced by a well-judged contrast.

By way of illustrating this intimate connection between floral compositions and the vessels containing them may be given the following ten artistic virtues attributed to certain special combinations:—

Simplicity;—expressed by Rushes and Irises in a two-storey bamboo vase.

Aspiration;—denoted by a vessel of decayed wood containing a climbing creeper.

Affection;—the character attributed to a bronze basin containing a Pine branch entwined by a Wistaria.

Serenity;—expressed by a suspended bronze boat, bearing white Chrysanthemums, suggestive of a loaded ship in port.

Austerity;—the sentiment conveyed by the *Eularia japonica* and *Patrinia scabiosæfolia* arranged together in a small bronze vase.

Quaintness;—denoted by a hooked vessel in the shape of a gourd, containing small Chrysanthemums.

Brightness;—the idea suggested by a bronze vase engraved with a design of wild geese flying across the full moon, and holding Lespedeza flowers.

Chastity;—the character expressed by a bronze vase engraved with a design representing rain, and containing a branch of Maple.

Security;—denoted by some kind of water plant, placed in a bronze vase engraved with the design of a spider's web.

Veneration;—the sentiment conveyed by a branch of Pine, or some other evergreen, placed in a bronze vase engraved with the representation of a crane. The crane and pine tree are both associated in Japan with the idea of venerable old age.

The above combinations, capricious as some of them may appear, serve to illustrate the manner in which vessels and flowers are used together to express an appropriate sentiment.

Sometimes the harmonious connection between the two is based merely upon a resemblance in the name of both. The Clematis, for example, is called *Tessen* (*Tetsu-sen*) and because the word *Tetsu* signifies *Iron*, this flower is often placed in a rough iron vessel. (see Plate XLVIIIb).

The native name for the Wistaria is *Fuji,* and the bell-like ornaments hung to the eaves of temples being called *Furin,* by way partly of a play upon the words, the Wistaria is sometimes arranged in an inverted bronze bell. (see Fig. 14).

Certain writers go so far as to classify flower vessels according to the seasons, recommending, for Spring arrangements, bamboo vases, bronze vases, and narrow necked vessels; for Summer compositions, flower-baskets, bronze basins, wooden tubs, or other broad-mouthed vessels; for Autumn designs, boat-vases, and porcelain vessels; and for those of Winter, gourd-shaped vases, and narrow necked vessels. However rare and valuable a receptacle may be, it

Fig. 14.

must not be used for holding flowers unless intended for that purpose; jars, jugs, pots, and other utensils having special uses of their own should not be employed for floral arrangements. This rule is apparently violated in the case of such important flower-vessels as the *Horse-tub*, *Well-bucket*, and *Flower-boat*. But these are exceptions which custom has sanctioned, and their names have reference rather to the original models from which they are copied. In each case, moreover, there is a special connection in idea between these receptacles and the flowers placed in them, so that the result has no element of incongruity.

In such matters, however, considerable license is allowed to masters proficient in the art, and especially in the case of floral designs for Tea Rooms, where the employment of curious vessels of all kinds is permitted. Some of these, such as the *Paste-pot*, *Octopus-pot*, and others, are illustrated in Plate XIXn.

A few general directions are laid down as to the particular kind of receptacles suited to certain flowers.

For flowers of large blossom, such as the Peony, the Chinese-basket is preferred, the Peony being considered the principal flower of China. Moreover, these large baskets are in character well suited to show off the ponderous blossoms of this plant.

For most water plants, low basin-like vessels, or vases with very broad mouths, are best suited, but the Narcissus requires a narrow-necked vessel to show it off to advantage.

For plants of short and stunted growth, having large leaves, a tub-shaped vessel is chosen; and for the Wistaria, Lespedeza, and *Kerria Japonica* some kind of suspended receptacle is preferred.

Other general directions are given as to the style of arrangement suited to special kinds of vessels.

In *Hooked-vases* the floral design should suggest plants hanging over a cliff, and must be arranged in the horizontal-triangle style, with or without a *Streamer*. In vases of this class, having a mouth at the side, the lines of the flowers must not cut the edge of the aperture.

In *Standing-vases* of bamboo, with two openings, the upper mouth should hold a *tree* and the lower one a *plant*, in accordance with natural scenery in which the tree branches occupy a higher position than the plants.

Often the same flower is used in both mouths of the same vessel, in which case some such distinction as the following exists:—Supposing Pine branches to be arranged in both openings, a style called the *Hill and Valley Pines* is adopted, in which the top branch represents the Pine trees on the summit, and the lower branch those at the base of the hill. The idea of distance must be suggested in the former, and that of proximity in the latter.

If the bamboo vase have one top opening and two side apertures, a composition called the *Hill, Plain, and Water* style is followed. To convey this idea a mountain *tree* is placed in the top, a land *plant* in the middle, and a water *plant* in the bottom opening.

FIG. 15.

Sometimes an arrangement exactly the reverse of this is devised in order to express the notion of a distant landscape with a mountain lake above, fields on the hill slopes, and a forest at the base. In such a composition, the uppermost mouth of the vessel contains a water *plant*, a land *plant* is placed in the middle aperture, and a *tree* branch occupies the lowest position. The *tree* branch should be arranged high to preserve the idea of a near foreground, the land *plant*, expressing middle distance, may be of moderate proportions, and the water *plant* at the top must be kept small and cramped in character to suggest distance.

In arranging flowers in tall bamboo vases of several mouths, the composition in the upper openings should assume the form of the horizontal or leaning triangle, whilst that in the lowest mouth should be treated in the style of the vertical triangle. (see Fig. 15).

In the case of a pair of *Well-buckets*, the upper vessel should have a *tree* and the lower one a *plant*. The rope should be kept sprinkled with spray as if covered with dew. Neither of the compositions in the two *Well-buckets* must be allowed to cross the line of the rope or chain which suspends them. Instead of the ordinary flower fasteners, *Well-buckets* generally have a perforated lid or frame which is fitted in the top just above the surface of the water. When these vessels are arranged in combination with a *Well-frame*, the upper bucket is suspended and the lower one rests on the edge of the frame, and in such a case the lower vessel may show water, but the upper one must preserve the idea of an empty bucket, and by no means must the water it holds be visible. In Spring time the floral design in the upper bucket should be the fuller of the two, but in Summer time the lower one should contain the more crowded arrangement. In Autumn both compositions should be simple and quiet. For arrangements in double *Well-buckets* the lower floral design should be of the style used for standing vases, and the upper one of the style employed for suspended vessels, with a lateral lean and a *Streamer*. Flowers arranged in square buckets should never be placed exactly in the angles or corners of such vessels.

In the kind of double bamboo vase called the *Row of Piles*, the higher tube should contain a land *plant* and the lower tube a water *plant*.

In the *Bridge-newel-vase*, a cylindrical vessel with a top mouth and an oblong side opening, the top of the vase should have a thick stump or heavy arrangement of *tree* branches, and the side aperture should contain some simple *plant*, modestly arranged.

In *Flower-baskets*, those with arched handles should have the flowers arranged so as to keep within the enclosure of the handle and not cross it. In very elaborate compositions this rule is sometimes violated, but in such cases the cutting or crossing must only take place on one side, and by no means in the centre of this side. The handleless baskets are generally hooked vessels, and flower arrangements in them are disposed as for this class of receptacles.

In *Sand-basins*, if a *tree* be used, it must be supported by a *plant* of some kind. Plants alone may be combined, but the composition must in such a case be full and strong to give it an appearance of solidity. Trees or plants arranged in these vessels are often divided into separate groups divided by a space (see Plates XXII and XXXVIII). If trees are used, the interval between each clump is technically called the *Valley-space*, and it should be equal to about one tenth the height of the arrangement. If water plants are used in divided groups, the distance between them is regulated in the same proportion, but different terms are used to denote this spacing. Supposing such water plants to be arranged side by side and in a line, the composition is said to have the *Fish-swimming* style, but if the flowers are placed one in front of the other they are said to be composed in the *Fish-sporting* style. By a curious fancy an analogy is here drawn between the relative position of the plants in such broad vessels and that of fish either swimming or sporting in a lake or stream. The arrangements of trees or plants in water basins, whether in single or double groups, must be always more towards one side of the vessel than the other, all formal and symmetrical compositions being disliked.

In *Horse-tubs* the employment of *tree* branches is prohibited, and *plants* of one or two kinds must be used. With regard to spacing, the same rules apply as are given for *Sand-basins*.

If a pair of similar vessels containing floral compositions be used side by side, the flower arrangement in one should be as nearly as possible the reverse in form of that in the other, but the colours of the flowers should be varied. For example, one vase may contain a red and the other a white flower, with the stems or branches of both disposed in nearly similar lines. These symmetrical arrangements are not, however, often resorted to.

In *Flower chariots* it is usual to arrange the seven flowers of Autumn; with these are sometimes combined other Autumn grasses, making the number up to nine or eleven different kinds of plants. Such elaborate combinations, which are not generally allowed in other flower vessels, are technically called *Embroidery*.

WATER IN VASES.

Various rules are observed as to the use of water in flower vessels. In Spring and Autumn the vase should be about nine tenths filled with water, in early Summer nearly full, and during the hottest days of Summer brim-full to overflowing, the rim of the vessel being oiled so that the water actually seems to pile up above the edge. In Winter time the vase should only be four fifths full, or as little as seven tenths in the coldest season.

When water plants and grasses are arranged in broad shallow vessels, the water forms part of the composition, and the different portions of its surface are accordingly regarded as having different properties. The water nearest the flowers is supposed to be moving and life-giving, and must therefore be free from floating matter; the portion removed from the flowers is considered stagnant and may contain floating weed or leaves.

To add more to the fresh appearance of floral compositions it is customary to sprinkle the outside of vessels with spray, conveying the idea of dew. Metal vases should not be wetted, but nearly all pottery or porcelain vases are so treated after the flower arrangement has been made. The Celadon vase is said to form an exception to this rule, because it becomes naturally covered with moisture.

Fig. 16.

It often happens that a vase filled with water, but without any floral composition, is used in the alcove of a chamber. Such an arrangement is resorted to when there is a picture on the wall of the recess representing some flower of the season, in which case a few petals of the flower represented in the painting may be put into the water of the vase. A vessel containing water only is an appropriate ornament for the chamber at a Moon-viewing party, or when a picture of the full moon is displayed, the intention being to suggest the idea of the moon reflected in a lake. Sometimes a few maple leaves are placed in the vase in order further to sustain the idea of a natural sheet of water.

STONES AND ROCKS.

In addition to the flowers, vessels, and fasteners, all of which contribute to the completion of a Japanese floral design, stones or rocks are sometimes added to arrangements of water plants and play an important part in the compositions. These stones represent in some cases the large bowlders which form stepping stones over streams and lakelets, in other instances they are meant to suggest islands in extensive water scenery. Again it sometimes happens that land and water plants are used in combination in a large shallow vessel and then the stones are disposed so as to suggest the dry bed or the banks of an adjacent river. Both white and black stones are employed, the white ones being placed near to the flowers, and the dark ones in parts of the water where there are no flowers. The flowers are arranged in front of or behind the stones and must not appear to grow out of them.

The chief ornamental stones in a flower basin are generally distributed according to the favourite triple principle which is applied also to the lines of the flower composition, under the distinguishing designation of Heaven, Earth, and Mankind, conveying in the present case the idea of elevation, horizontality, and intermediate height. One stone is of vertical character and supposed to resemble a mountain, one of flat and nearly horizontal character, and the third, which is placed between the other two, partakes of an intermediary character. Other stones of secondary importance are added to set off the larger stones and generally to connect the whole composition.

The use of such stones may be seen in Fig. 23, where they are arranged in a large basin together with Wistaria flowers. In Fig. 16, on the previous page, is shown a fancy arrangement in which blocks of charcoal are employed instead of stones.

FLOWER TRAYS AND STANDS.

Fig. 17.

All standing flower vessels, with the exception of the *Flower-baskets*, are placed upon a square tablet of polished or lacquered wood interposed as a protection between them and the surface of the dais or shelf on which they stand. A story is related elsewhere which accounts for the absence of such tablets under standing flower-baskets, but a very reasonable explanation seems to be that these baskets cannot themselves stain or damp the dais, as they contain an inner vessel which holds the water and the flower stems. Even such a simple object as the vase tray or tablet has its fixed measurements, and a few fancy forms in the shape of fans, circles, or segments are not uncommon. Sometimes this flat tray is replaced by a small ornamented stand or table of carved or lacquered wood, intended to raise and give more importance to small flower arrangements. As previously stated, excepting in the case of the low tub-like vessels, intended to suggest water scenery, and used for water plants, the surface of the water and the fastener holding the stems in the vase should be a little above the eye level of the seated spectator, so as not to be visible without effort.

Some flower stands exist which are of considerable height, having a bottom shelf. These are used for Incense Meetings, in which case the top of the table carries an ornamental incense burner, the shelf below being occupied by a very simple flower arrangement (see Fig. 18). Examples of different trays and stands are shewn in Plate XXb.

PLATE XXII.

Blitum Beetroot in Basin with Flowers as Fontaine.

Begonia Everniana in Basin with "Rush Fontaine."

Narcissus in Basin with "Branched Fontaine."

Scolpus lacustris in Basin with "Reed and Cord Fontaine."

Kerria japonica in "Tobacco-Tub" with "Horse-Hair Fontaine."

Rosa rugosa in Basin with "Paper String Fontaine."

Iris in "Well-Spring Vase with "Crab Fontaine."

POSITION OF FLOWERS IN ROOMS.

ALL important rooms in a Japanese dwelling, large or small, are provided with an ornamental alcove or recess called the *Toko-no-ma*. In chambers of a superior class this recess is of grand proportions, occupying the right hand half of the side wall, and is furnished with a raised floor, polished and lacquered, and a handsome corner pillar of rare wood. The remainder of this wall, to the left, is occupied by a corresponding recess fitted up with ornamental shelves and cupboards, collectively called the *Chigai-dana*, on account of their picturesque irregularity of arrangement. The division between these two alcoves is marked by an ornamental pillar in front, and is sometimes filled in with a thin plastered partition pierced with a window-like opening. If the *Toko-no-ma* is to the left and the *Chigai-dana* to the right, this characterizes the room as one of secondary importance. In some chambers the *Toko-no-ma* is used alone, without the adjoining shelves, in which case it appears to project into the room, the corner pillar standing forward. It is this recess in all Japanese dwellings which contains the principal moveable decorations, and to which attention is mainly directed as the point of interest in the room. Here the householder displays one by one his choicest art treasures, and it is in front of this alcove that the most honored guest is placed.

The back wall of the *Toko-no-ma* is the surface upon which the *Kakemonos* or rolled pictures are exhibited. These paintings are hung singly, in pairs, in triplets, and occasionally in fours. The floral design is placed upon the dais of the recess, below and to the front of the pictures, unless a hanging arrangement of flowers be used, in which case it is suspended from the lintel or ceiling, or hooked to the pillar of the alcove.

It is considered important that the floral composition should not in any way clash with the pictorial arrangement, either as regards position, line, subject, or sentiment. The two together must form a harmonious decorative composition.

When only one wall-painting is exhibited, the vase of flowers in front should be placed, as a general rule, rather to one side; if two pictures are hung, a single floral design should occupy the interval between them; in the case of three pictures, two vases of flowers should be used, one opposite to each interval; and with four paintings, three flower compositions are employed in a similar manner. In the last case, instead of using three floral arrangements, the central space is sometimes occupied by a statuette or incense-burner.

Vases containing flowers are often elevated upon a small raised stand or table instead of the flat board or tablet which is placed under most flower vessels. One kind of table used has a shelf below, on which a very simple floral design may be arranged, and when so employed the top of the table supports some other ornament.

The proportions of the *Kakemono* or hanging picture influence the disposition of the flowers on the dais. In front of a long painting the floral composition must be kept as low as possible, but when a short and broad picture is displayed the flowers may stand high and full in arrangement. The object of this rule is to prevent the mural painting from being hidden by the floral design. The same result is often obtained by placing the vase of flowers to one side of the recess, instead of in the centre. It is sometimes unavoidable that the flowers cover part of the picture, but under these circumstances special care must be taken not to hide that portion bearing the stamp and signature of the artist. The centre, ends, and tassels of the ornamental roller forming the bottom border of the painting must also be always unobstructed. When the pictorial work contains figures, the features of these figures must on no account be hidden by the branches of the flower arrangement, and if, as occasionally happens, the picture is inscribed with a poem or proverb, this writing must not be obscured.

It is moreover important that the flowers should accord in character and sentiment with the picture before which they stand. Occasionally this principle is carried so far that the floral group becomes, as it were, a continuation of the scene delineated in the painting, but when this result is aimed at, great care should be taken that the one does not detract from the value or excellence of the other.

PLATE XXVII.

JAPANESE INTERIOR WITH ARRANGEMENT OF SPRING FLOWERS.

Thus, if the picture represents lake or river scenery, it is preferable to adopt a floral composition consisting of water plants, of the kind that might actually exist in the foreground of such a landscape. It is, however, necessary to avoid using in the vase flowers of the same kind as any depicted in the painting, because such a juxtaposition would inevitably tend to detract from the excellence of the picture, which must suffer by a close comparison with nature. For example, supposing a picture of blossoming Plum trees to be hung to the wall of the recess, the vase on the dais in front should not contain Plum blossoms but some other seasonable flowers. Under special circumstances courtesy might demand the violation of this rule, in which case the floral arrangement should be rough and unassuming, so as to detract as little as possible from the skill of the painter. In front of a painting representing some flower of the season it is not unusual to place a vase of water containing a few fallen blossoms and petals of the flower represented; in this way the idea of the picture is continued, without occasioning invidious comparisons.

As explained in a previous part of this work, an important distinction is observed between flowering *trees* and *plants*. The rough and bold character of the former, in comparison with the softer and more graceful nature of the latter, renders a judicious combination of the two, in floral arrangements, productive of pleasing variety. The same principle of combination is applied to the connected arrangement of floral design and wall picture; and it is considered preferable, if the painting displayed represents flowering *plants*, that branches of *trees* should be used in the flower composition, and *vice versa*.

Again, if the hanging decoration be a Chinese manuscript, and not a painting, the arrangement of flowers in front must be simple and modest in style; and if the work of a very famous caligrapher, poet, or painter, be exhibited, it is often best to have no floral composition at all. In ordinary cases, if a verse of poetry be displayed in the recess, the flowers may assist to illustrate the poem; thus, with verses descriptive of the beauties of Chrysanthemums or Pine trees, the flowers of the Chrysanthemum or branches of Pine should be employed in the vases in front. Some Schools lay down a rule exactly the reverse of this, holding that natural flowers should not be placed before a sonnet extolling their beauties, as they are apt to restrict the imagination and detract from the pastoral sentiment conveyed in the poem.

Reference has been made elsewhere to the double associations from animal and vegetable life, used as favourite art motives by the Japanese. Such combinations are;—Bamboos and Sparrows, Lions and Peonies, Nightingales and Plum blossoms, Deer among Maples, Wild-horses amid flowering grasses, and many others that might be mentioned. A combination of flowers and picture so as to produce a connected composition of this sort is considered very desirable. According to such a method of combination a picture of Deer requires in front of it an arrangement of Maples, a painting of Horses needs wild flowers, one of Lions necessitates the use of Peonies, and representations of Dragons demand Pine branches for the floral design in the foreground. In the same way with figure paintings, when the figures represented are traditionally associated with particular trees or flowers, such flora should if possible be employed for the flower arrangements used before them. When, for example, a picture of Hotei, one of the Gods of Fortune, is displayed, Bamboo branches should be disposed in front, and before pictures of Chinese children, a common subject with Japanese painters, coloured flowers are appropriate.

The connection of idea between the wall painting and the floral composition is occasionally one based upon the reputation of the painter or upon some fiction with which his name is associated. A famous Chinese painter called *To-emmei*, whose works are greatly valued in this country, is said to have professed a great passion for Chrysanthemums, hence, when a painting by this artist is displayed, it is customary to use Chrysanthemums in the flower arrangement. Plum blossoms are reported to have been the special fancy of another great painter— *Kin-nasei*, and these flowers are therefore placed before his pictures.

This kind of combination is quite irrespective of the subject of the paintings except in cases where other important rules would be violated. If the particular flowers required are represented in the paintings, it would then be an error to use the same natural flowers in front. Such a selection would not only be redundant, but would tend to detract from the excellence of the painting. It must also be remembered that the pictorial hangings in a Japanese room are frequently changed and are not displayed continuously throughout the year as is the custom with oil paintings in European rooms. They are quite as much an expression of the season and occasion as the floral compositions themselves. The contingency, therefore, of pictures requiring flowers which are out of season, and not obtainable, could hardly occur. To return, by way of explanation, to one of the examples just quoted, it would be obviously incongruous to place Plum blossoms, the flowers of early Spring, before a painting by *Kin-nasei*, if such picture represented an Autumn landscape; but such a subject would not be displayed during the season of Plum blossoms, because it would be considered out of place as a decoration for that time of year. In fact, one of the charms of Japanese mural decoration is this constant harmony between the art motives employed and the particular period of the year. One is not called upon to study snow scenes in Summer or mountain torrents in Winter time, for the pictorial design which occupies the place of honor in the chamber is always in tune with the present, and conveys a sentiment in harmony with the special season and occasion.

The floral compositions as well as the wall pictures must accord in character with the chamber itself. Thus, for a superior room of handsome dimensions, the *Kakemono* should be the work of a famous painter or caligrapher, and the flowers should be full and showy in arrangement. The work of a young painter of little repute, or a floral design of a very simple or rough character would be considered out of place in such important rooms. But for a chamber of secondary dimensions and importance, rare pictures and elaborate floral compositions would be inappropriate. In very small rooms it is better entirely to dispense with a painting when flowers are exhibited, supplying the place of the picture by a simple tablet.

Besides the above orthodox methods of combining pictorial hangings and flower decorations, other fancy combinations are sometimes resorted to. In a large chamber recess, suited for two or three paintings, when one picture only is exhibited, there is a manner of using flowers, and at the same time expressing the idea of three pictures, by hanging on either side of the painting vases of flowers attached to mural tablets. The flowers thus employed must not be the same as any represented in the central painting. A similar treatment is followed with two pictures, by placing one hooked floral design in the centre. Such arrangements are considered more perfect if they form one connected idea with the picture or pictures. Taking the example of two paintings, one of a Tiger and the other of a Dragon, a vase of Bamboo branches might be hung in the centre, because the Tiger is always represented in Japanese pictures as roaming in Bamboo groves.

Reference has been made before to the ornamental tier of three shelves in a Japanese chamber, placed at the side of the principal recess or *Toko-no-ma.* Important floral designs should occupy the dais of the *Toko-no-ma*, but for small and secondary arrangements the *Chigai-dana* or *Irregular Shelves* are often employed. As already explained in the case of the three openings of a high triple-mouthed vase, so for this tier of three shelves, triple arrangements of flowers should follow the natural distribution of growth observable in real scenery. For the top shelf, thick and moss-covered *tree* branches are chosen, for the middle shelf young *tree* branches or land *plants* are selected, and for the lowest shelf a composition with water *plants* should be employed. If *tree* branches be placed on the middle shelf, then the lowest shelf may have land *plants* instead of water *plants*. In the same recess, and in combination with the shelves, there is occasionally constructed an ornamental cupboard about eighteen inches or two feet in height. The paper slides of such cupboards are often painted with flower designs, and, in this case, care must be taken not to use natural flowers of the same kind in vases placed on the adjacent shelves.

Strict rules are established as to the exact position in the *Toko-no-ma* or principal recess to be given to suspended vessels. First, with regard to *Boat-shaped* vessels, it is stated that in ancient times they were hung at a distance of about three feet from the ceiling or lintel, and exactly in the centre of the recess. In later times it became customary, however, to suspend them, at a distance equal to about a quarter of the span of the recess, from the corner pillar, which is always on the guests' side of the room, being that nearest the light. If, as is not uncommon, two *Boats* are suspended together, the upper one should be about sixteen inches below the lintel, and the lower one about the same distance above the floor of the recess; but this height is sometimes changed in order to suit the wall picture.

The direction given to the prows of such vessels, so as to suggest different ideas of motion, has been already explained. *Boats* are in some cases hung to the under sides of ornamental shelves or cupboards.

Well-buckets are not considered suitable for the recesses of important rooms, unless they happen to be a present from a superior, in which case they may be given the place of honor; they are, however, frequently resorted to in second class rooms. When one bucket only is used, it is placed standing upon the floor of the recess, raised about fifteen inches from the surface by a stand of some sort, and removed about eleven inches from the corner pillar. Square buckets must be placed angle-wise. In the case of two suspended buckets, supposing the height from floor to lintel of the recess to be divided into three parts, the top of the lower bucket should be one third and the top of the upper bucket two thirds of this height, from the dais. The bottom of the lower bucket thus becomes about eighteen inches from the floor of the recess, and is supported upon a small table, or the edge of a stand representing a well-frame. The position of the well-pulley will be about one third of the span of the recess from the corner pillar, and the rope of the lower bucket being inclined and tort it will be brought well to the side of the recess. If, as is usually the case, flat-sided buckets are used, one will have its sides parallel to the wall and the other will present its angle. Occasionally the above mentioned proportion as to height is violated, and the lower bucket is placed immediately upon the floor, with the interposition only of a board. A style exists called the *Mossy-spring-buckets* in which this board is sprinkled with stones and moss. Sometimes a slab of decayed wood, or an irregular row of bamboo tubes resembling a Japanese drain-board, are used under the lower bucket. The *Piled buckets*, consisting of two standing buckets, one supported on the edge of the other, are placed upon a flat drain-board immediately on the floor of the recess, and near to the corner pillar. One of these should be parallel to the wall and the other placed diagonally.

Hitherto flower compositions have been considered with reference to their disposition in the permanent recesses of chambers, either in the *Toko-no-ma*, or on the fixed shelves of the *Chigai-dana*. Other fancy arrangements, having no connection with the chamber recesses, also exist, though they are rare. Among such may be mentioned the arrangement of floral designs upon *Cabinets*, and *Flower-horses*.

The *Flower-horse* is so named because it is made in imitation of the ornamental clothes-horse or frame which is used in Japanese sleeping apartments, either for hanging the garments upon, or for carrying large strips of rich curtaining serving as screens. The construction of such *Flower-horses* consists of two vertical and two horizontal bars of lacquered wood, framed together with the top bar projecting at the ends, the bottom bar being

FIG. 18.

steadied and supported upon short cross-pieces. Such frames measure about five feet square. In the example given in Plate XXVI, vessels of different kinds containing a variety of floral arrangements are hung to the side posts and cross-piece. From the centre is suspended a crescent-shaped bronze vessel containing an arrangement of small Chrysanthemums and a trailing Vine; from the right hand pillar are hung a

cylindrical bamboo vase, with side mouth containing a drooping arrangement of
Patrinia scabiosæfolia, and, below, a globular basket with a composition of Carnations
combined with the branch of some tree; and to the left hand pillar are fastened a
hooked bronze vase with *Papaver rhœas*, and, below, a horn-shaped bamboo vase
containing a double arrangement of *Platycodon grandiflorum* with Barley.

There exist also fancy kinds of flower-stands consisting of the *Flower-horse*
combined with shelves and cabinets, and on which standing, hooked, and suspended
flower compositions are arranged.

Flower-cabinets, in their simplest form, consist of two small shelves of different heights connected by vertical and horizontal framing, lacquered black, and ornamented with metal. For the upper shelf of such cabinets a drooping floral design
is generally adopted, and for the lower shelf a standing arrangement. In the example
shewn in Plate XXVI, the top shelf supports a bronze vase with trailing Ivy, and
the lower shelf a porcelain vase containing Asters. The same illustration shows a
fancy cabinet called the *Thatched-kiosk*. It consist of a bottom board, with raised
shelf supported on bamboo posts, and covered by a rustic thatch-roof. The shelf
carries a bamboo vase from which springs a creeper, arranged to pass over the roof
and hang down the side as a *Streamer*.

Other articles of furniture are sometimes pressed into service for the display
of flowers. An example exists of a triple gong-frame, with one upper and two
lower openings, from the cross bars of which cylindrical bamboo vases containing
different kinds of Peach blossom are shown suspended.

A great fancy prevails for collecting and displaying together different kinds
of blossom of certain favourite trees and plants. The use of distinct varieties of the
same blossom in one arrangement is opposed to the somewhat austere rules of the art
of floral composition, but such specimens are sometimes displayed in separate vessels
arranged on stands. The famous regent Hideyoshi is said to have devised an
arrangement of five baskets of green bamboo disposed on the two shelves of a
Flower Stand and containing different varieties of the Cherry blossom. Such arrangements, however, belong more to gardening than to chamber decoration.

Some writers go so far as to say that the floral design in a chamber should
have a contrast in style with that of the adjoining garden. This fancy is better

appreciated if it be remembered that during a great part of the year the outer walls of the house, which consist almost entirely of paper slides, are thrown completely open.   If there be a landscape garden adjoining, consisting of lakes and hills, the floral arrangement in the room should by preference partake of a moorland character; but if the garden be level and waterless, then water plants or mountain trees should be selected for the flower decorations of the chamber interior.

PLATE XXVIII.

JAPANESE INTERIOR WITH ARRANGEMENT OF SUMMER FLOWERS.

CEREMONIAL AND ETIQUETTE.

THE Art of arranging flowers in Japan is essentially a polite accomplishment, and is governed by important rules and restrictions as regards etiquette and ceremonial. It is presumed that all floral designs are made mainly with the object of giving pleasure to visitors; and on certain ceremonial occasions they are actually intended to convey a silent compliment to the principal guest. Receptions given in rooms where flowers are arranged often partake of the character of *Flower Meetings*, the guests in turn inspecting and admiring the host's floral compositions, or being called upon by him to make arrangements of their own.

Fig. 19.

In attending such a reception, the visitor should leave his fan in the ante-chamber and, approaching within about three feet of the recess, seat himself in the old ceremonial attitude with his knees bent and the body resting back on the heels. One hand should be placed on the knees, while the other respectfully touches the mats in front, the body being slightly bent forward. It must be remembered that there is always a supposed connection between the pictures, which adorn the wall surface of the recess, and the floral arrangement standing or hanging in front. The guest should therefore first regard the *Kakemono* or picture, and if, as is often the case, there are three of these, he should examine first the central, then the left hand, and lastly the right hand one. Having thus bestowed his admiration upon the background of the scene, he may slide a little closer and inspect the floral composition in the foreground. In doing so he should first observe the central line of the flower arrangement, and then gradually examine left and right, and from top to bottom of the composition; lastly, with a word of apology for so doing, he may inspect the manner in which the stems are held at the bottom, this being one of the most difficult parts of the *technique*. It is considered impolite to put the face behind the branches and peer too closely into the flowers. After such inspection the guest slides a little back and regards the whole composition from a respectful distance, using some suitable expressions of admiration. Rules of etiquette actually go so far as to give the exact words to be employed in admiring different designs; for it is considered bad taste to apply indiscriminately exaggerated terms of praise not in keeping with the character of the particular flowers. When inspecting hanging floral arrangements the seated attitude should be changed for a standing and stooping posture.

A visitor is often invited to make an extemporary arrangement of flowers, for which purpose he is presented with certain suitable flower stems, or blossom-clad branches, and all the necessary utensils and implements. On such occasions the

FIG. 20.

host must provide a vase, three quarters filled with water, which should be placed in the centre of the recess upon a tray or table spread with a sheet of paper. In addition to this, a flower tray with two **or** three kinds of cut flowers, just as they were gathered, with withered leaves and dead twigs all left intact, a pair of scissors, a knife, a small saw, and a folded flower cloth or duster, must be placed on the dais to the left of the vase, or in some convenient position on the floor of the chamber. The length, width, and manner of folding the flower cloth are all prescribed. Near to the above utensils and implements must be placed a water jug full of water, and several forked twigs suitable for *Flower-fasteners*. These various tools and utensils are illustrated in Fig. 20.

After asking a guest to arrange flowers, the host should commence to roll up the *Kakemono* which adorns the recess, as it is considered to be demanding too much of a visitor to expect him to extemporize a flower arrangement in harmony with the picture which happens to decorate the recess at the time. The guest may, however, prevent the painting from being removed; thereby tacitly undertaking to make his flower composition accord with it. Should the master of the house produce a very rare and valuable vessel for holding the floral arrangement, it is polite for the guest to make objections, pleading want of sufficient skill to do justice to so precious a receptacle. If pressed, however, he must attempt a simple and unassuming arrangement of flowers, so as not to detract from the merit of the vessel itself. Should the host produce an insufficient quantity of cut flowers, the guest must make the best of them and on no account ask for more.

The visitor who is about to make a floral composition approaches the recess in which the flower vase is placed, and seats himself sideways towards it, facing the light, which in chambers of a superior kind is on the left side. He then fixes the flowers as quickly as possible, changing his position in order to regard them from a point immediately facing the recess, and altering and correcting them from this position. If the arrangement is intended to have a connection of idea with the hanging picture, it is placed fronting the edge of this picture, on the side nearest the light; but if no connection is intended, then it may be disposed centrally. Under no circumstances must the floral design project beyond the corner pillar of the recess. Having completed the composition, the designer should, as a matter of compliment, ask his entertainer to fill up the vase with water, and if this request be declined, he may then replenish it himself; he should not, however, press the host on this point, because a correct apportionment of the amount of water suited for different arrangements requires considerable knowledge of the flower art, and it is possible that the host may decline on account of ignorance. When the water is filled in, the stand of the flower vase is wiped, and the different implements, with the exception of the scissors, are all put on the tray, and placed near the serving entrance or right hand side of the chamber. The scissors are purposely left near the flowers as a silent and modest invitation to the master of the house to correct faults. The host brings the dust pan and brush, sweeps up any fallen leaves or litter, and removes the tray of tools.

If the hanging picture has been removed during the arranging of the flowers, the guest must now re-hang it, and see that the floral composition is placed so as not to clash with it in any way. When the whole arrangement is completed, the host and any other visitors present, who have meanwhile remained in an adjoining

room, enter, and approaching in turn the *Toko-no-ma* or recess in which the flowers
are placed, salute and inspect in the manner previously described. The master of the
house naturally confines his admiration to the floral arrangement, but the guests
should also find compliments for the receptacle in which it is placed. After the
Flower Meeting is concluded and the visitors retire, the guest who has arranged the
flowers should remove them from the vase, placing them upon the board on which
the vase stands, or on the wash-basin of the adjoining verandah; unless he is specially
requested to let his work remain, it is considered presumptuous for him to quit
without destroying the evidence of his skill. This rule does not apply to visitors
of very superior rank, who may be considered to show honour to their host by
leaving their designs intact.

Politeness during such meetings is considered so important that, if a rule
happens to be violated through ignorance or incapacity on the part of the entertainer,
the guest must do his best under the
circumstances, and must do nothing to call
attention to the error. As an example may
be given the instance of a host producing
scented flowers for a visitor to arrange at
an *Incense Meeting*, which is an occasion on
which all flowers with perfume are prohibited.
The guest must in such a case use the
flowers, removing the full-blown blossoms, and
making an arrangement of buds alone in as
simple and unassuming a manner as possible.

In giving presents of cut flowers
for the purpose of flower arrangements, they
must not be trimmed, or they will look as if
they had been previously used. The sender
must, however, consider how they are capable
of combination into a floral composition, and
must include stems and branches suitable for
the accessory parts of the design. Such
flowers should have the bottom of their stems
placed in fancy paper wrappers. The form
of paper wrapper suited to *tree* branches and

FIG. 11.

that used for *plants* differs slightly. When flowers are offered as presents the number of buds should exceed that of the open flowers, and they should be accompanied with withered or worm-eaten leaves, and cobwebs, all left intact so as to look as if the flowers were freshly gathered. The recipient of such presents should carefully consider how the cuttings can best be arranged without injury or extensive alteration. If they appear to him quite unsuitable for a proper floral composition, it is better to place them in a vase as they are than to attempt a formal arrangement. In Fig. 21 is shown a branch of Plum blossoms held in a paper wrapper.

Presents of flowers are sometimes made to superiors in a kind of hand-bucket, which should be of white wood and quite new. Several sorts of flowers are generally placed in such vessels, which are furnished with lids perforated with four or eight square holes for the purpose. When such a present is received, it may be put on one of the shelves of a recess in a reception room, an empty vase being placed on the dais.

In a previous part of this work flowers considered specially felicitous during particular months have been enumerated. It remains, however, to consider, under the head of Ceremonial, certain floral arrangements which are fixed for important festivals. Of the numerous fête-days celebrated in Japan, may be mentioned first that of the New Year, and then the five great festivals called *Go-sekku*. The *Go-sekku* occur in all the months of the year of odd number, with the exception of the eleventh month, and in four cases out of the five fall on the day corresponding to the number of the month.

FLOWERS FOR THE NEW YEAR.

For the floral decorations of the New Year it is customary to adopt a combined arrangement of Pine, Bamboo, and Plum branches in a large bronze or porcelain vase. These *flowers* are however sometimes used separately, in which case the Pine is displayed on the first, the Bamboo on the second, and the Plum on the third day of the year. In some cases a vase of green bamboo, with twigs and leaves left on, is used to hold branches of Pine and Plum trees, the floral triad being formed by including the vase itself. The Willow is a favourite tree for use in hanging compositions at this season; and plants such as the *Adonis amurensis* (Fukuju-so), *Rhodea japonica* (Omoto), and *Ardisia japonica* (Yabu-koji), are often employed in combination with the Plum.

FLOWERS FOR THE "FIVE FESTIVALS."

The seventh day of the New Year is the first of the Five Festivals (*Go-sekku*) and is called the Fête of the Seven-grasses (*Nana-kusa*). This festival is of Chinese origin, and its meaning is involved in some mystery, but it is undoubtedly connected with ideas of luck in the number seven. At its celebration seven different herbs are struck with seven different household tools, and divine protection is invoked against the evils of the year. The flower arrangements adopted on this occasion do not differ from those used for the first three days of the New Year.

The third day of the third month is the second of the Five Festivals, and is called the Fête of the Dolls. It is the national fête for girls, which they celebrate by displaying richly attired images representing the Emperor and Empress surrounded by courtiers. On this occasion Peach blossoms, Willow branches, Cherry blossoms, the *Rhodea japonica*, and *Rosa indica*, are chosen for flower designs.

The third of the Five Festivals, falling on the fifth day of the fifth month, is the Fête for Boys, called, by the Japanese, *Tango*. On this day large painted paper fish made hollow so as to fill and float with the wind are placed on high bamboo poles in front of the dwellings, together with standards and other ornamental signs, each male child of a household being entitled to display one fish. The flower most honoured on this occasion is the Flag or Iris. A kind of early Chrysanthemum, and a particular sort of Bamboo, called the *Moso-chiku*, are also used.

The next festival falls on the seventh day of the seventh month and is called the Fête of *Tanabata*. It is the celebration of the meeting of two stars, and is supposed to be a day of very good omen. Slips of paper inscribed with poems and leaves of bamboo are hung up as ornaments on the occasion, and a general feasting takes place. The flowers used for compositions on this holiday should be *Platycodon grandiflorum* (Kikio), Bamboo, and *Anthriscus arguens* (Karukaya), arranged together, or *Patrinia scabiosæfolia* (Ominaeshi) employed singly. A special arrangement consisting of a couple of suspended bronze boats, called the *Facing Boats*, because their prows face, and containing paper Chrysanthemums, is used. The paper flowers should be of seven colours, and among them should be one real flower. To the *Streamer* of the floral design in one boat should be hung two, and to that of the other, five loops of silken cord of five different colours,—white, yellow, blue, purple, and red. These cords are called the *Worshipping-cords*, being associated with the idea of worship to the stars. A somewhat similar method of composition is adopted for standing vases, seven different flowers being placed in three vessels. The central vase contains three flowers, of which the *Principal* should be formed of a branch of the sacred tree *Cleyera japonica* (Sakaki), and the other two vases should each hold two flowers. The floral designs in all three vases must have their stems bound with the *Worshipping-cord.*

The last of the Five Festivals occurs on the ninth day of the ninth month, and is called the Fête of Chrysanthemums. Its origin is Chinese, and it is connected with a fancy that this flower imparts long life to those who imbibe its blossoms in wine. For this occasion, as the name of the festival implies, the Chrysanthemum is invariably used for floral arrangements. Flowers of five colours are employed for first-class compositions, White for the *Principal*, Light Red for the *Secondary*, Dark Red for the *Tertiary*, Blue or Purple for the *Sub-principal*, and Yellow for the *Support*. Sometimes a design made with flowers of one colour only is fancifully called the *Three colour* composition, the green of the leaves and the black of the bronze vase being included in the colours. The *Rhodea japonica* (Omoto), and *Nandina domestica* (Nanten), are also sometimes displayed on this occasion.

In addition to the Five Festivals just enumerated there are other felicitous days in the year on which special rules are adopted for the flower arrangements.

For the first day of the sixth month, which is the beginning of the *Dog-days* or hottest season, water plants should be arranged in a sand basin or broad-mouthed vessel, together with white sand and pebbles, which are supposed to suggest ice.

For the first day of the eighth month, called the *Hassaku*, the flowers used should have fruit or berries, and all faded leaves or branches must be carefully removed. This fête was one specially honoured by Iyeyasu, the first of the Tokugawa Shoguns, to celebrate his entrance into Yedo.

On the eighth day of the eighth month white flowers should be displayed.

For the fifteenth day of the eighth month, flowers of the season should be employed, with the addition of a sheaf of the ripe rice-plant in celebration of the harvest.

The above special rules, though hardly applicable to any European adaptation of the art under consideration, are interesting as showing how a certain harmony of idea should always be sought between the floral design and the special occasion celebrated. The following are ceremonies which may occur at any period of the year, with general rules for their appropriate flower arrangements.

FLOWERS USED AT BETROTHALS.

On this occasion visits take place and presents are interchanged. The flower or flowers specially felicitous to the particular season, in accordance with the list previously given, must be used. The Iris is considered specially suited for this purpose, unless out of season. Some people, however, object to the use of flowers of Purple colour.

FLOWERS FOR WEDDING FESTIVITIES.

With regard to flowers arranged for weddings, it must be remembered that, amongst colours, Red is regarded as *male* and White as *female*. Hence in the case of a son-in-law being adopted into the family of his bride, the bridegroom being regarded as the guest of the occasion, the *Principal* line of the floral design should be Red, whilst the *Supporting* line is of the *female* colour—White. On the other hand, when a bride is adopted into the family of her husband, she being considered the guest of the occasion, the White colour has the central position in the arrangement. In both cases the stems of the flowers used must be closed, and firmly connected at the base with coloured cords, called *Mizuhiki*, to signify union. The Bamboo and Pine, being always obtainable and specially felicitous, are considered well suited for wedding ceremonies. They should be placed separately in a pair of similar vases, the Pine towards the guest's side of the chamber, and the Bamboo towards the host's side. Purple flowers, Willow branches, and other drooping plants are prohibited for weddings, as also are all floral compositions placed in suspended vessels. Purple is considered expressive of mourning, and all loose drooping arrangements imply want of union and constancy.

FLOWERS FOR COMING OF AGE CELEBRATIONS.

There were formerly two coming of age celebrations for male children, one on the occasion of the boy first assuming the ceremonial trousers, and the second when the long locks were cut off and the youth assumed the cue worn by adults. For both of the above festivals, flower arrangements were required to be firm and vigorous, with a large proportion of buds and young branches. Faded branches and full blown flowers were prohibited.

FLOWERS FOR PROMOTIONS IN RANK.

Flowers used on the occasion of honourable promotions should have buds below and open flowers above, to signify ascent in rank, and the use of withered leaves or over blown flowers must be carefully avoided.

FLOWERS FOR THE CEREMONY OF RELIGIOUS RETIREMENT.

Upon reaching a ripe age it has always been the custom in Japan to seek retirement from active life, the head being shaven and a semi-religious ceremony performed. Floral designs used at such a celebration should consist of berry-bearing plants or trees, and red flowers must not be employed.

FLOWERS FOR OLD AGE CELEBRATIONS.

Old moss-clad branches, flowers late in season, and second flowerings are suited to such occasions. Withered branches and flowers which quickly fade or fall to pieces must on no account be used.

FLOWERS AT FAREWELL GATHERINGS.

There are certain flowers which bloom twice in the same year, and these are technically called *Returning-flowers*. Though considered unsuitable for ordinary occasions, because out of harmony with the season, such flowers are specially appropriate for farewell gatherings. The idea in so employing them is to express the hope for a safe return. Similar arrangements were adopted at feasts before a battle or campaign.

FLOWERS FOR THE SICK.

Flowers placed before sick persons should be put together in a rapid and unlaboured manner, and should be vigorous in style, to suggest the idea of recovery and strength. At the ceremony of praying for the sick, flower compositions should be full and gay, as well as bold and powerful in style. The use of the Pine, *Podocarpus macrophylla*, and *Rhodea japonica* should be avoided.

FLOWERS BEFORE HOUSEHOLD SHRINES.

Each household in Japan has generally two shrines, one dedicated to the household gods of the old Shinto cult—the national deities of the country, and the other to the spirits of deceased relatives, which is Buddhist. For arrangements of flowers before the Shinto shrine a full and powerful composition is required. All ugly flowers, those of strong odour, and those having thorns are prohibited. A special branch called the *Facing-branch* is used behind the *Principal* line of the composition and pointing towards the shrine. The floral design placed in front of a Buddhist shrine should also be full and crowded, but the arrangement should not be laboured. The front of the design must face the shrine and not the spectator. A story is told of a Daimio who went to his ancestral mortuary chapel and arranged flowers with great skill and labour much to the admiration of the chief priest, who, however, pointedly asked him whether he had made his composition with a desire for public praise or in reverence for the spirits of the dead. Unable to answer, and struck with the justice of the reproof, the noble altered the flower arrangement to one more simple and unassuming. White flowers are specially suited for such designs, and in all cases as little artificial trimming and bending as possible should be resorted to, the withered leaves only being removed.

FLOWERS AT DEATH ANNIVERSARIES.

For floral designs from the first to the fiftieth anniversary of a death, a quiet and simple style must be adopted. White and Yellow flowers are used, and with them is combined a branch of the sacred tree *Cleyera japonica* in token of a religious offering. Sometimes a withered branch is employed for the *Principal*. As such flower arrangements must be unassuming, and the composer must not attempt to display skill, it is considered ignorant and in bad taste to criticise them, judging them by the ordinary standards. The selection of flower buds is to be avoided, full-blown or over-blown flowers being preferred. The use of crossing leaves, or flowers blooming twice in the year is also forbidden.

On and after the fiftieth anniversary gayer arrangements and even red flowers may be employed, and the *Principal* line should be formed with an old moss-covered branch, a flower of the season being added as an auxiliary. Flowers of ominous names, such as the *Oni-yuri*, (Demon-lily) must on no account be used.

FLOWERS AT PRAYERS FOR RAIN OR FINE WEATHER.

There are special rules with regard to flowers which are suited for the occasions of prayer for rain or fine weather. As it is the East wind which brings rain, floral arrangements used at time of praying for rain should have their *Principal* line pointing from right to left, to suggest the East wind blowing. A reverse arrangement is resorted to on occasions of prayer for fine weather, when the *Principal* line leaning from left to right is made to suggest the West wind.

FLOWERS USED AT MOON VIEWING.

Moon viewing is at all times a favourite pastime of the Japanese, but the great moon festival of the year is on the fifteenth day of the eighth month. The more important dwellings have a special chamber with open galleries from which the sight of the moon-lit landscape can be enjoyed. The floral arrangement occupies the recess of the chamber and has of course no real connection with the outside prospect, but in the flower composition itself the idea of a moon-lit landscape is expressed. A branch of a Pine tree is used, and between the *Principal* and *Secondary* lines of the composition a special branch is introduced, fancifully called the *Moon-shadow-branch;* a hollow gap also is formed between the foliage, bounded by a special branch called the *Dividing-branch.* In the composition the idea is to suggest both the opening through which the moon can be partially observed, and the dark branch which appears to cross its surface. To fully appreciate the analogy one must be familiar with the scenery of Japan, and have seen the irregular Pine trees standing out against the starry heavens.

FLOWERS FOR HOUSE WARMINGS.

The Japanese word for Red (*Hi*) is phonetically the same as that which denotes fire, and as conflagration is the dreaded enemy of the beautiful but perishable buildings of the country, a superstitious objection exists to the use of red flowers on the occasion of such celebrations as house-warmings. Flowers whose names contain the ominous ideograph *Hi* are also prohibited. Such flowers are, the *Helianthus* (Himawari), *Lychnis grandiflora* (Gambi), *Lilium concolor* (Himeyuri), and the *Thuya obtusa* (Hinoki). The use of the last, (Hinoki), is specially disliked, as the wood of this tree was anciently employed for producing fire. By a curious perversion however, some Schools approve and even favour the use of *Hinoki*, because the name is phonetically synonymous with another word *hinoki* meaning protection from fire.

For floral designs on occasions of this kind a two-story bamboo standing-vase is suited, having water plants arranged in both openings, the reason given for such an arrangement being that water is the extinguisher of fire. Another favourite combination in Summer is that of a white Magnolia branch with white Irises. In Winter, suitable flowers are scarce, and the Narcissus is used in the lower opening of the vase, with water alone in the top.

FLOWERS AT POETRY MEETINGS.

A favourite pastime of the educated classes in Japan in former times, was the meeting of friends for the composition of verses. Sometimes a flower of the season formed the subject of competitive versification, and in this case the Enshiu School held that the flower chosen as a theme, if in season, should be displayed in a simple and quiet arrangement in the chamber of meeting. Some Schools were, however, of a contrary way of thinking, maintaining that the particular flowers selected as subjects for poetical composition should not be displayed at the meeting.

FLOWERS FOR THE TEA CEREMONIAL.

As the chambers in which the Tea ceremonial is conducted are of very limited size, the flower arrangements must be small in scale, simple, and unaffected.

PLATE XXIX.

JAPANESE TEA ROOM WITH ARRANGEMENT OF AUTUMN FLOWERS.

Plate XXIX illustrates a Tea Room with the guests coming in through the low opening that forms the entrance to these fairy chambers, and having a simple arrangement of Autumn flowers hooked to the wall of the alcove. The displaying of the floral design has its proper order in the ceremony. It is not generally placed in the recess at the same time as the hanging picture. First, some suitable painting is shown, and a part of the entertainment takes place, after which the guests retire; the *Kakemono* is then removed, and the flower arrangement is prepared before their re-admission. Partly to avoid keeping the visitors waiting too long, and partly on account of the severe simplicity of the surroundings, and the austere character of the entertainment, the flower composition is made as unassuming as possible. The kind of floral design adopted for Tea Rooms, differs therefore from the ordinary style, being less elaborate and studied, and is distinguished by the name of the *Thrown-in-style*. If a standing vase be employed, the composition consists of the *Principal* line used singly; in a hooked vase, the *Secondary* line alone is used; and for a suspended vessel, the *Tertiary* line is preferred. If three stems are combined in one design, they must not be divided, but must be kept together in one line. Flowers used on such occasions must be well sprinkled with spray, to look fresh and as if covered with dew. When a Tea ceremony takes place in a large room, instead of the orthodox miniature chamber, then both the flowers and mural picture may be displayed at one time. If such meetings are held at night, floral arrangements are not employed. A hooked or suspended arrangement of flowers is preferred for Tea Rooms. Red blossoms and flowers having strong perfumes are disliked. The following flowers are also objected to:—*Patrinia scabiosæfolia*, Cherry, Rose, Lotus, *Cnicus*, *Calendula officinalis*, *Nuphar japonicum*, *Celosia argentea*, *Illicium religiosum*, Orchid, Pomegranate, Daphne, *Lespedeza*, Drooping Lily, Drooping *Platycodon*, Tea Plant, and *Anthistiria arguens*.

One of the objects sought in arranging flowers in a Tea Room is to arouse the admiration of the guests by the quaintness and novelty as well as the unassuming simplicity of the composition. For this reason curious and unusual vessels are often resorted to.

FLOWERS FOR INCENSE MEETINGS.

Another polite pastime of the Japanese in former times was that of burning different kinds of incense, and of distinguishing the scents produced; a small room like the Tea Room being generally used on such occasions. The little bronze or porcelain incense-burner became the principal art object of the chamber, and was supported on a stand or low table, about one foot high, placed on the dais of the recess. The height of this stand necessitated that the hanging picture should be a broad low one, technically called a *Cross-picture* (Yoko-mono). The most simple kind of floral decoration at these entertainments was that of a single flower and leaf, or a small bunch of flowers, placed in a tiny vase below the stand or table (see Fig. 17). Some theorists object to this arrangement because the incense-burner above contains fire, and both wind and fire are injurious to flowers. If not used in this manner, the flowers should be placed in a vase hooked to the pillar or wall of the recess, or a suspended boat arrangement may be adopted. All scented flowers of whatever kind are prohibited at incense-burning meetings. This rule excludes such flowers as the Plum, Daphne, Chrysanthemum, and *Aconitum fischeri.*

In cases where a hanging incense-burner is suspended from the lintel of the alcove, a standing arrangement of flowers may be placed below.

PLATE XXX.

JAPANESE INTERIOR WITH ARRANGEMENT OF WINTER FLOWERS.

PRACTICAL EXAMPLES.

THE following illustrations, from Plate XXXI to Plate XXXIII inclusive, are given to aid the designer by showing in juxtaposition, defective and corrected arrangements of the same floral branches or leaves.

In Plate XXXI the *Kerria japonica* is represented at A, placed in a standing vase, with its branches very much in the form assumed when first gathered. The defects of this composition are, the stiffness of the central stem, the weak and open appearance of the branches at the base, the redundancy and parallelism of some of the smaller sprays, and the general fulness, irregularity, and formlessness of the whole. Altered as in B, the central stem or *Principal* assumes a stronger curve, the different branches are united at the base, the lines of the *Secondary* and *Tertiary* are improved, and some of the smaller sprays removed, so that the whole becomes disposed into five governing lines.

In Plate XXXII the *Spiræa cantoniensis* is shown at A, carelessly arranged in a hooked bamboo vase, having a side mouth. Here the central line is too long for a hanging arrangement, which requires the drooping branch or *Streamer* to be the longest, and the branches are too numerous and straggling. Altered as at B, the *Principal* is shortened, the *Secondary*, being in this case a *Streamer*, is bent down in a wave-like curve, some of the excessive sprays are removed, and the whole resolves itself into a three-lined arrangement.

In Plate XXXIII the Leaf-orchid is shown at A in a defective and almost symmetrical arrangement. The principal faults of this composition, are the straightness of the central leaf and the distribution of the fronts and backs of the leaves, as shown by shading, which is monotonous and incorrect. Improved, as at B, the stem of the central leaf is gracefully curved, showing its front surface with the edge curled over, and the other leaves are united firmly at their base, arranged in steps, and disposed so as to reveal front and back surfaces in a well balanced variety. The whole assumes the character of a trilineal design.

The special rules and traditions which govern the arrangement of particular flowers will now be explained, practical examples being shown, so far as the scope of this work permits, in accompanying illustrations.

ARRANGEMENT OF PLUM BRANCHES.

The Plum blossom being the earliest flower of the year, is held in high esteem for floral arrangements. The hardiness of the Plum tree, the duration of its blossom, its sweet perfume, as well as the austere type of its beauty, all help to make it even a greater favourite for flower compositions than its more showy rival the Cherry tree. There exist in Japan many varieties of this blossom, red, white, single, and double. Branches of red Plum blossoms should be arranged in a fuller and gayer manner than those of white blossom, which should be displayed in a quiet and open style. For Spring arrangements a fancy prevails of suggesting, by means of a perch-like bend of a branch in the composition, the presence of the Nightingale, which delights to make its home of song in the Plum trees. Heavy and antique standing vases are the most suitable receptacles for this flower.

The rough and irregular character of the branches of this tree, renders considerable trimming and bending necessary in order to arrange them into lineal compositions, and this manipulation is assisted by softening refractory portions in hot water. In Plate XXXIV is illustrated the process of composing a trilineal design with small branches, each stem being shown both in its natural and its altered form, side by side with the completed design.

But the favourite kind of Plum branch for chamber decoration is that of the thick, rugged, moss-covered trunk, with young branches and shoots attached. The fresh shoots, which grow vertically in parallel clusters, are specially admired in combination with the rugged trunk, and are introduced under a special name (*Zuwaye*) in an appropriate place in compositions. Arrangements of this kind vary according to the class of room in which they are used. For important chambers a composition leaning to the right is adopted, and for secondary rooms one leaning to the left.

A   B

UNSTUDIED AND DEFECTIVE ARRANGEMENT.   ALTERED AND CORRECT ARRANGEMENT.

B

ALTERED AND CORRECT ARRANGEMENT

A

UNSTUDIED AND DEFECTIVE ARRANGEMENT

A — DEFECTIVE ARRANGEMENT.

B — ALTERED AND CORRECT ARRANGEMENT.

Such designs are said to have respectively South and North character. For arrangements of *Right* or *South* character, full and half-open blossoms predominate, buds being employed sparingly in the *Tertiary* branch only; whilst for those of *Left* or *North* character no full blossoms are used, but only buds and half-open flowers, a more quiet and bare character being given to the whole composition. The shoots are, in both designs, sparingly introduced, springing from the main trunk in three parallel lines, and in a position between the *Secondary* and *Tertiary*. In this particular, the Plum tree forms a privileged exception to the general rule which prohibits *Parallelism* in the disposition of stems or branches.

A curious arrangement of Plum branches in a large shallow basin filled with water, illustrated in Plate XXXVA, requires special notice. As a general rule these broad water-basins are employed only during Summer, and for the display of trees or plants which grow near or in water. In the present case, the composition consists of an irregularly bent cutting of a Plum tree disposed in an oblong shallow vessel, in the leaning style, with the principal branch diving through the water and re-appearing beyond. The explanation of this fanciful arrangement is that it was originally in imitation of the famous Recumbent-Dragon-Plum-Tree at Kameido, remarkable for its crawling trunk which ploughed the ground in several places before re-appearing blossom-clad beyond. In the artificial composition it is considered important that the extremity issuing from the water should rise firmly, to suggest springing from the earth, and should, in fact, be fixed as if it were a branch separately held below, though sufficiently conveying the idea of continuity of line with the original branch. Some Schools otherwise explain the origin of this style of design. By them it is called the Plum of the Mountain Stream, or the Water-diving Plum, and its introduction as a floral design is attributed to the æsthete Soho, who, during a hunting expedition, observed a crooked old Plum tree diving in this manner in a mountain stream, and adopted the idea for application in the flower art which he practised. A tree of this kind measuring over a hundred feet from its root to the extremity of its branches is said to have existed near Kioto. In the above arrangements, sand and gravel are mostly used for fastening the Plum branch in position, but sometimes the *Horse's-bit fastener* is employed; in which case, for reasons already stated, the composition must not be placed on the dais of a chamber of importance. In one of the illustrations of chamber interiors (Plate XXVII) is shown a composition with Plum branches in a broad shallow vessel as above described. The clothing of the seated figures and the snowy landscape indicate the early Japanese Spring,— the season of the Plum blossom.

The Plum is used in combination with land and water plants such as the Spring Chrysanthemum and Narcissus (see Plate XXXVIa). An elaborate composition in a high bamboo vase with five mouths is sometimes seen, having a branch of white Plum in the top, a Willow branch below, then a Narcissus, after which a Camellia, and lowest of all a Pine branch. In Plate XXXVii may be seen a branch of Plum blossoms arranged in a hanging basket, in combination with the plant *Adonis amurensis*.

ARRANGEMENTS OF PINE BRANCHES.

The Pine is the most important of all flowerless trees in Japan. The hardiest and noblest of evergreens, and a constant feature of the landscape,—whether it be mountain or coast scenery,—bent and twisted by the wind into shapes so quaint and contorted that faithful imitations in miniature may well be mistaken for grotesque caricatures, this venerable tree forms a favourite subject for poets and artists, and finds an important place in floral compositions on all congratulatory occasions. Combined with the plum and bamboo, and associated with the crane and tortoise, it is used in decoration to express the sentiment of happy old age.

The principal kinds of Pine are the *Pinus thunbergii*, known by the Japanese as the *black* or *male* Pine (Kuro-matsu and O-matsu), the *Pinus densiflora*, called the *red* or *female* Pine (Aka-matsu and Me-matsu), and the *Pinus parviflora* (Goyo-no-matsu). The *Pinus parviflora*, on account of the straightness and delicacy of its leaves is often arranged in a simple vertical style, using thin sprays, but for compositions with other species of this tree, thick gnarled branches are preferred, and a bold rugged character given to the design. The Pine is used mostly for arrangements in standing vessels for which it is said to be specially suited, but hanging compositions are occasionally employed. A favourite treatment is that of a broad stump cut off horizontally, with a thick twisted branch springing from its base. At all times the Pine branch is used as much as possible in its natural state, being trimmed until its arched masses of foliage assume a balance suggestive of the three radical divisions employed in flower compositions, and the building together of separate branches, common with other trees, is comparatively little resorted to. An example illustrating the Pine foliage disposed in cloud-shaped masses may be seen in Fig. 3, page 55.

The occasions for which the use of the Pine is specially appropriate are at the New Year, at weddings, at old-age celebrations, and sometimes, though rarely, at farewell gatherings. At wedding feasts a double arrangement in a pair of similar standing-vases is employed. For this purpose a branch of the *male* Pine is placed in one vessel, and a branch of the *female* Pine in the other. The general form of each design should be similar, but the branch of the *female* Pine which faces the opposite vase should stretch a little beneath the corresponding branch of the *male* Pine. These together are called the *Destiny-uniting* branches, and the complete design typifies eternal union. The same sentiment is expressed by arranging a branch of each of these trees, one below the other, in a single vase.

Occasionally in suspended arrangements of Pine, long stiff threads are hung from the branches, in conventional imitation of the parasitic grasses which attach themselves to this tree; and in disposing such threads, their balance into groups of three, five, or seven irregular lengths is carefully attended to.

The white Chrysanthemum is sometimes used in combination with the Pine, a custom introduced by the philosopher Rikiu in allusion to a favourite verse of poetry comparing the white flowers of the wild Chrysanthemum when seen beneath the Pine branches to the moon between black clouds,—the foliated leaf-masses of the Japanese Pine tree, piled one above the other, being not unlike in shape to rolling masses of dark clouds. This is one of the many examples in which the sentiment of landscape is introduced into flower designs. Another example is that of what is called the *Mountain-valley-and-stream* composition, suited to a vase of three openings. In the upper mouth is placed a Pine branch, to suggest a mountain summit; a land plant is placed in the centre, to suggest the plain or valley; and in the lowest mouth is arranged a water plant, to convey the idea of a stream or river. High bamboo vases are employed for such arrangements. Somewhat similar to the last named composition, is a design in a vase of two mouths with Pine branches in each, the top branch being kept small in order to suggest a distant tree on the hill-top, and the lower branch full and elaborate to represent a tree in the foreground.

The use of the Pine at moon-viewing gatherings has been alluded to in an earlier part of this work. Its employment in a single arrangement at farewell-gatherings was invented by Rikiu, being suggested to him by a verse of poetry in which the Japanese name for Pine (Matsu) conveys a double meaning, the same word also implying *waiting* for a lover's return. Many of the traditions of the Flower Art are based upon such poetical allusions.

Examples, though rare, may be seen of the Pine arranged in a water-basin in combination with a water-plant, such as the Iris or Narcissus. One form of fancy arrangement called the *Fuji-pine* after the famous volcanic mountain Fuji-san, remains to be mentioned. In such a composition a bent branch is selected to resemble the outline of mount Fuji, and is combined with the other branches and foliage in such a manner as to give the profile of the bare conical peak and suggest the wooded country at its base.

ARRANGEMENT OF BAMBOOS.

In the Flower Art, the Bamboo is, strictly speaking, regarded neither as a *tree* nor a *plant*, but it may occupy the position of either. It should never be placed in a vessel made of bamboo. In combination with other trees or plants thin branches or sprays of Bamboo are often employed, but for simple compositions a portion of the round stem or tube is selected, with a few leaf-clad twigs attached. The top of such tubes are cut off either in a splayed or horizontal manner according to the occasion. If used at wedding feasts, the cutting must be hidden by leaves, as the sight of it is considered to be unlucky and suggestive of severed friendship. These cylinders of green bamboo require very special and careful treatment, in order to preserve their verdure and vitality for any length of time. They must be cut in the cool of the morning or evening. In the early months of Summer, when the leaves are young, the stem is very succulent, and no special treatment is required, but to preserve the freshness at other seasons, small holes are drilled between the knots of the tube, into which water is blown with the mouth. Sometimes *sake*, the rice-wine of the country, is added to the water as a stimulant.

Mostly three, but sometimes five twigs of leaves are left on these cylindrical stems, arranged irregularly. The leaves of such branches are in their natural state crowded and confused, and they require thinning out, the withered ones being removed, and the remainder disposed in double or triple groups. Three different combinations of leaves are approved, that of the *Fish-tail*, in which two of the lancet-shaped leaves spread out like the tail of a common fish; that of the *Gold-fish-tail* in which a central leaf is added, giving the group the resemblance to the triple-finned tail of a gold fish; and that called the *Flying-goose* shape, consisting of three sloping leaves, suggestive of the form of a wild goose in flight.

B

COMPLETED ARRANGEMENT.

A

NATURAL AND ALTERED FORMS OF BRANCHES FOR "PRINCIPAL,"
"SECONDARY" AND "TERTIARY" LINES.

PLATE XXXV.

PLATE XXXVIII.

A

B

On some occasions, two tubes of green bamboo, one shorter than the other, are combined, called respectively the *male* and *female*. The top of the *female* or lower tube should be cut off horizontally, while that of the upper tube should have a splayed cutting. The longer tube has three knots and two leaf-clad twigs which take the place of *Principal* and *Secondary* in the composition, and the shorter one, only two knots and one spray of leaves corresponding to the *Tertiary*. To designs placed in broad vessels a Bamboo sprout (Take-no-ko) if in season, may be added as an auxiliary, and it should be fixed about two inches distant from the main stems, generally in a slanting position and against an ornamental stone.

Numerous vases made of bamboo cylinders have been described in an earlier part of this work. Similar vessels are often made of green bamboo, the leaves being left on, and forming part of the floral composition arranged in such vases. The same care is required to keep such receptacles fresh and green as is applied to ordinary bamboo tubes, and while the upper and lower portions are cut out to form mouths for the purpose of receiving other plants, the intermediate space between two knots is bored and sprinkled internally with water and *sake*. In Plate XXXVIII at A may be seen an arrangement of Bamboo in a sand-basin together with the *Nuphar japonicum*, and at B a design with thin stems of bamboo alone.

COMBINATION OF PINE, BAMBOO, AND PLUM.

Separate notice is given to the combined arrangement of Pine, Bamboo, and Plum branches, on account of the importance attached to such triple compositions in the Art under consideration. This combination, called by the Japanese *Sho-chiku-bai*, is used at important celebrations in token of congratulation and well-wishing. It is specially employed at the New Year, and, if the Plum blossom be in season, at wedding feasts. Sometimes the composition is in a single vessel, and sometimes in three. If all three be placed in one vase, the Pine takes the position of the *Principal*, the Bamboo of the *Secondary*, and the Plum of the *Tertiary;* and if arranged separately, the Pine occupies the central position, the Bamboo is placed on the left, and the Plum on the right. The Plum branch should have its base tied with a coloured silken cord, when used at weddings. A bamboo vase is not considered suitable for such triple arrangements, but a vessel of green bamboo may be employed with the Pine and Plum only, thus including the vase itself in the combination. An illustration of this is shown in Plate XXXIXa, and another example of the *Sho-chiku-bai* may be seen in the same plate at B.

ARRANGEMENT OF WILLOW BRANCHES.

The Willow is employed for floral designs on important occasions from the fifteenth day of the eighth month to the first day of the third month. Owing to the facility with which its branches can be bent, it is considered the easiest of all *trees* to arrange after the Japanese manner, and generally forms the subject of first lessons in the Art. Among the many varieties of the Willow found in Japan, those most used for floral compositions are the *Salix purpurea*, or ordinary river-side Willow and the *Salix babylonica*, or Weeping Willow. When branches of the Weeping Willow are employed, care must be taken to avoid an arrangement in which lines droop on both sides of the same composition, such designs being only permitted at the celebration of death anniversaries. It is a common practice to tie the long trailing stems of the Willow into a loose loop or ring of about four inches in diameter, generally three stems being so united, and the extremities hanging down at different lengths. The originator of the Enshiu School is said to have invented this style of composition, as he found it otherwise difficult when arranging Willow branches in standing vessels, to prevent the long shoots from trailing on the floor. Another version attributes the practice of looping the Willow to an old Chinese custom prevailing at farewell meetings, on which occasions it signified tying-up until the return of the departing guest. Hence it is said to be a style of arrangement specially suited for farewell gatherings.

The custom became afterwards applied even to hanging compositions, which are generally preferred for this tree, the idea being that, because the Willow tree grows near water, its branches should hang over laterally as if drooping over a stream. Even in chambers of handsome dimensions, for which standing flower arrangements are mostly selected, the Willow is frequently placed in a vase hooked to the recess-pillar, or in boat-shaped vessels suspended from the cross-beams. A design of a Willow branch in combination with a bunch of Narcissus placed in a hooked basket may be seen in one of the text illustrations Fig. 8.

The Camellia is the favourite flower for connecting with the Willow, a

B

A

A

B

combination at variance with the general rule that two *trees* ought not to be employed together in double compositions. At the season when the buds of the Willow present a reddish appearance the white Camellia is chosen, but when these buds assume a greyish white colour the red Camellia is considered more suitable. In Plate XLII is shown an arrangement of Willow and Camellia in a suspended bronze vessel of crescent shape, and in Plate XL are illustrated two compositions, one consisting of the Willow combined with the *Camellia sasanqua*, and another of the Willow and Narcissus. If both are arranged in a two-staged bamboo vase, the Willow should be above, and the Camellia below. The Narcissus and Winter Chrysanthemum are also occasionally used in combination with the Willow.

Three fancy styles of composition are recognized for the branches of this tree, namely, the Willow in fair weather, the Willow in wind, and the Willow in snow. For the fair-weather style the Spring Willow is used, and the branches are spread as if just kept apart by the breeze; when the Willow in wind is expressed, the branches are given a curved sweep, as if blown back by a strong wind (see Fig. 22); and when the Willow in snow is suggested, the stems are made to hang straight and heavily as if weighted with snow.

Fig. 22.

ARRANGEMENT OF WISTARIA FLOWERS.

The Wistaria, as the first flower of Summer and the most important of blossoming creepers, holds high rank, though its colour renders it unsuitable for certain occasions,—purple being associated with mourning. The species with white blossom is seldom used in floral designs. From its nature this plant is specially adapted to suspended arrangements, though it is also used in standing vessels; in the latter case some sort of frame is generally necessary for its support. The blossom-clusters are preferred before they are in full bloom, consisting of a majority of buds and half-open flowers, with only three or four fully open.

Tubs and sand basins, such as water plants are placed in, are the only kinds of standing vessels used for the Wistaria, which grows in parks and gardens on trellises overhanging a lake or stream. The idea of such water-scenery is conveyed by means of these broad flat vessels, and stones are often introduced to add to the impression of landscape in miniature.

An arrangement with ornamental stones is shown in the accompanying woodcut (Fig. 23). Such stones are generally three in number, with three distinct characters distinguished by the names Heaven, Earth, and Mankind. The principal stone is high and pointed, like a mountain in shape; the third stone is flat and horizontal in character; and the second stone is of intermediary form. Other stones of secondary importance are often added, but not so as to detract from the three essential ones, and the complete number should always be an odd one. The stem of the Wistaria should spring from behind the principal stone. Water plants like the Iris and *Nuphar japonicum*, and land plants such as the *Calendula officinalis*, Bijin-so, and *Aspidistra lurida*, are sometimes combined with the Wistaria in shallow vessels. If land plants are introduced, no stones should be employed, these being only resorted to when water scenery is expressed.

Fig. 23.

A composition with Wistaria, Irises, and *Nuphar japonicum* would be arranged as follows:—The Wistaria should be placed to the left of the vessel at the side of a high stone,—as if it were on the slope of a hill with its branches stretching over and its flowers reflected in an adjoining lake or stream. It should have five, seven, or nine clusters of blossoms. Then, at a distance of about six inches, a group of the leaves of the *Nuphar japonicum* is placed; and four and a half inches from this plant are arranged the Irises in a simple composition of three or five leaves, one of them curling over and dipping into the water. The remaining stones should then be distributed in different parts of the vessel, but so as

B

A

not to interfere with the water plants, which must be placed in the free water-space.

The nature of the Wistaria-creeper prevents its stems being arranged in distinct lines in accordance with the general methods of floral design, but the rules for *Principal, Secondary* and *Tertiary* are applied to the grouping of its blossom-clusters. In hanging arrangements the *Streamer* assumes great importance.

Examples of suspended compositions may be seen in Plate XLIIA, where the flower is placed in a bronze boat-shape vessel, and in Fig. 14, page 76, where it is arranged in an inverted bell twining round a branch of Pine.

ARRANGEMENT OF IRISES.

Several kinds of Irises are employed for floral arrangements, the most important being two varieties of *Iris lævigata* (Kakitsubata and Hanashobu), the *Iris sibirica* (Ayame), the *Iris tectorum* (Ippatsu), and the *Iris japonica* (Shaga). Of these, the *Iris lævigata Fisch. var. Kæmpferi*, which the Japanese call Kakitsubata, is the favourite. This plant, though belonging particularly to the early Summer, lasts through several seasons, displaying a special character of growth at these different periods. In Spring, the leaves are stiff and straight, and the flower stems are short; in Summer, the leaves are more full and spreading and there is much spirit in the flowers; and in Autumn the flower stems are long and the leaves bent and curled. These peculiar characteristics of the plant at different seasons must be shown in the floral arrangements. The leaves of the Iris, like those of other water plants, are considered the most important part of the composition, and they must be well selected, all withered ones being discarded. They are plucked in their sheaths, separated, and then artificially connected into groups of two or three, being attached with saliva. These sets of leaves are used to form the different lines of the design, names being given to each according to its position and function in the composition. The main leaves form the *Principal, Secondary,* and *Tertiary* lines of the arrangement, to these being added two or three flowers and other auxiliary leaves. Plates XLIII and XLIV demonstrate the methods of taking to pieces the clumps of leaves, re-arranging them, and combining them, with the addition of flowers and buds, to form a simple composition.

The usual numerical proportions between leaves and flowers in an arrangement of Irises, are:—one flower with three leaves, two flowers with seven leaves, two flowers with fifteen leaves, three flowers with thirteen leaves, and five flowers with eleven leaves.

Taking, by way of example, a composition of two flowers and seven leaves, the following is the method of arranging. First, three leaves are placed in the position of the *Secondary* line. These three leaves are joined together with the two outer ones long, and the central one short, as if just sprouting out from between the other two. Above this group of leaves a full-blown flower is placed, with one long leaf added, which occupies the position of the *Principal*. This is often called the *Cap-leaf*, as it crowns the whole. Behind it a smaller leaf is added as an auxiliary, and at the side of this a flower bud is introduced. Next a small leaf is placed in the position of the *Tertiary*, technically called the *Water-dipping-leaf*, because its tip curls over to the water of the vase; and combined with this is added another small leaf called the *Dew-supporting-leaf*, because its blade should be slightly concave above with an upward tendency.

In making a composition of eleven leaves and five flowers, three leaves are put in for the *Secondary* as before, and a flower is added. Behind this, and rather higher up, the *Cap-leaf* is placed, and with it is arranged the highest flower as *Principal* with two other auxiliary leaves. Then another flower is added between the *Principal* and *Secondary* blossom with an additional leaf below it; and on the opposite side of the composition, a little lower than the flower of the *Secondary*, is placed another flower with two leaves adjoining. Still below this is added a bud, for the *Tertiary*, combined with two leaves,—the *Water-dipping-leaf* and the *Dew-supporting leaf*. Of the flowers mentioned, two should be in full bloom, two partly open, and one in bud.

Reference has been made to flower arrangements in broad flat vessels in which trees or plants are disposed side by side in divided clumps. The distance between such clumps in the case of *tree* arrangements, is called the *Valley-space*, but in that of *water plants* it is called the *Fish-travelling-distance*, because a lake or stream containing such plants is suggested, and the space left is just sufficient for the passage of large fish. Such interval should measure about one tenth of the height of the largest group of plants.

In a divided arrangement of Irises in two clumps, one should be large and

PLATE XLIII.

A

B

UNEQUAL SIZES OF LEAVES ARRANGED IN

ROSETTE COMPARISON OF STEMS.

LEAVES TOO LONG NEEDED FOR ARRANGING

FLOWER STEMS WITH SLIGHT DETECTIVE STEMS ARE NEW DEFECTIVE FLOWER
IMPERFECTIONS, ARRANGING

VIEW
MARKED

PLATE XLIV.

A

B

ARRANGEMENT IN A VASE WITH SINGLE FLOWERS.

NE-JIME STYLE WITH IRISES.

VARIOUS WAYS OF ARRANGEMENT IN CLASSES.

high, in the *Vertical-triangle* style (see Fig. 1, page 47), and the other small and leaning, in the *Horizontal-triangle* style (Fig. 2, page 48). The main group, placed to the right side of the basin, may have seven large leaves, a flower, and a bud; and the other group, six small curled leaves, and one bud.

The Japanese floral artist delights to suggest in his compositions some noted natural landscape. The most famous spot in Japan for displays of the Iris, is Yatsuhashi in the province of Mikawa, where there is a river with eight tributary streams crossed by as many bridges. An artificial flower arrangement is sometimes made in suggestion of this natural view, by using a very large shallow basin in which are placed white pebbles to represent the river and its eight branches, divided by black stones, filled in for the land.

The Irises are then arranged amongst the black stones, in divided groups. Each group is composed separately with a different quantity of leaves and a proportionate number of buds and flowers. Some bunches are disposed vertically, others in a more leaning style, and numerous pairs of young leaves are arranged between, to connect the whole. This arrangement is one example of a kind of flower composition almost resembling miniature gardening. The plants are held in their position by the sand and pebbles, and no other visible fasteners should be employed.

The *Iris sibirica* is arranged in a somewhat different manner, the flowers being placed higher than the leaves, whereas, with the *Iris lævigata*, the blades rise above the blossoms. Another peculiarity is that the leaves are arranged in regular gradations and not in indented triplets.

The Iris may be seen in Plate XLV. at A, arranged in a large basket together with a branch of Fir, and at B it is illustrated as an independent composition in a small bamboo vase. One of the subjects in Plate XXVA. is that of Flags placed in a *Well-frame* vessel and held by a metal *Crab-fastener*. In Plate XLVI. a large design of Irises is illustrated, forming one side of a paired floral arrangement, the *Aster tataricus* being placed in another vase opposite to it. The Iris is often arranged with other water-plants, such as the *Nuphar japonicum* and *Rhodea japonica*, in a broad vessel. It is also to be seen combined with trees, such as the Wistaria, Maple, and *Nandina domestica*.

ARRANGEMENT OF PEONIES.

Two distinct varieties of this plant are employed for floral designs, namely, the *Pæonia moutan* and the *Pæonia albiflora*. The former produces blossoms of immense size and is generally arranged in wide-mouthed bronze vases or antique flower-baskets, no other flowers being permitted in conjunction. The *Pæonia albiflora* —a smaller species which goes by the name of *Shaku-yaku*,—may be used in combination with branches of flowering *trees*, but should not be joined with other *plants*. The *Pæonia moutan*, or large Peony, is one of the three flowers to which are attributed royal rank, the other two being the Cherry and the Lotus. It must therefore be used alone and placed in the position of honour in a chamber,—that is, on the dais of the principal recess. It should never be arranged on a shelf or in any secondary place, and all other flowers must be excluded from the room which it is used to decorate. This rule is professedly followed out of respect for the Peony as queen of all flowering plants, but its luxurious character really renders such practice consistent with purely artistic considerations, the addition of other material being calculated to produce a surfeit of richness and elaboration. The employment of the Cherry blossom is also subjected to the same restrictions.

It is customary in arranging the large Peony to introduce between the principal stems one or two black withered twigs, which, by contrast, enhance the appearance of the leaves and flowers. As the blossoms are massive and heavy in character they should be sparingly introduced between the leaves, rarely more than one full-blown blossom being used; the remainder consists of buds and partially opened flowers. For the ordinary composition in three lines, five or seven blossoms may be employed, one for the *Principal*, one for the *Secondary*, one for the *Tertiary*, and the remainder distributed in intermediary positions. The leaves are carefully disposed so as to give *support* to the flowers, and in some cases are made to cross in front of them. The leaves surrounding the *Secondary* blossom should be large and closely grouped. They are technically called the *Lion-hiding* leaves, because the presence of the Lion,—which is associated with Peonies in decorative art motives,— is supposed to be suggested behind the thick foliage. A flower bud should be used for the *Tertiary*, with the addition of strong young leaves.

The *Pæonia albiflora*, or small species of Peony, is not held in so much honour as the larger kind, and is often combined with other plants. In the case of this plant, half opened blossoms are preferred for the *Principal*, full blossoms for the

PLATE XLV.

A.

B.

PLATE XLVI.

ASTER TATARICUS (SHIWON).

IRIS OF FIVE FLOWERS.

PLATE XLVII.

A

B

Secondary, and buds for the *Tertiary*. The Peony is sometimes arranged in wide basins divided into separate groups.

Illustrations of this flower arranged in different kinds of vases may be seen in Plate XLVII. at A and at B.

ARRANGEMENT OF LOTUS FLOWERS.

As has been stated in an early part of this work, the Lotus plant is associated with the spirits of the dead, and is therefore considered out of place as a decoration at festive gatherings. It, however, holds high rank in the Floral Art, being regarded as the king of Indian flowers. When employed on the dais it must be given precedence of all other flowers, and it would be considered a gross violation of taste to combine with it the Peony, which is the royal flower of China. In the case of the Lotus plant, the leaves play the most important part in the composition, this being a rule which applies to nearly all water-plants. The Lotus leaves should be selected to express the idea of the three Buddhist divisions of time—Present, Past, and Future. Past time is represented by a partly decayed or worm-eaten leaf; Present time, by a handsome open leaf,—often called the *Mirror-leaf*, on account of its resemblance in shape to that of a Japanese mirror ; and Future time, by a curled leaf, not fully open.

This plant should be arranged in a wide-mouthed vase, *Sand-basin*, or *Tub*, in which vessels it is often united with other water-plants. As an example of such compound arrangements may be described a combination with the Iris, *Nuphar japonicum* (Kohone), *Alisma plantago* (Omodaka), and Rush, with the addition of ornamental stones, so as to form a sort of miniature lake scene. A general rule applying to floral designs in broad basins, is that tall plants suggest shallow water, whilst those short in growth suggest deep water; and as the Lotus flourishes in comparatively deep water, the stems of its leaves, in a compound design, are kept short. The idea of *Principal*, *Secondary*, and *Tertiary* is followed in the disposition and character of the leaves; for the *Principal*, a large round leaf being employed, for the *Secondary*, a smaller leaf, and for the *Tertiary*, a young and curled leaf. An open blossom is introduced between the *Principal* and the *Secondary*, and a flower bud, kept low in position, is placed between the *Principal* and the *Tertiary*. The Irises are next arranged at a distance of about two and a half inches from the Lotuses, an open flower being used for the *Principal*, a half open flower with three leaves for the *Secondary*, and below, in the place of the *Tertiary*, two or three more small leaves. The *Nuphar japonicum* is next placed in position, its *Principal* consisting of a large oval leaf of interesting shape, and below it a half open flower occupying the place of the *Secondary*, whilst for the *Tertiary*, a young and only partly opened leaf is used. The Rushes are then arranged in five or seven bunches, and the *Alisma plantago* is afterwards introduced, separated from the Rushes by a short space, with two leaves having a bunch of flowers between them, the composition being similar in style to that of the *Nuphar japonicum*, but bent in an opposite direction. Of the five plants in this combination the Lotuses and Rushes are disposed in the *Vertical-triangle* style and the others in the *Horizontal-triangle* style. After they are all arranged in position, the stones are added to connect the whole group together.

ARRANGEMENT OF CHRYSANTHEMUMS.

Floral compositions with Chrysanthemums are considered very difficult to arrange skilfully. Of this flower there are many kinds, some of large and some of small blossom, the latter presenting least difficulty in treatment. There are varieties peculiar both to Summer and Autumn, and also a later species exists called the Winter Chrysanthemum. Characteristics of growth at the different seasons are faithfully imitated in floral compositions. In Fig. 24, are shown three separate sprays of the small Summer Chrysanthemum, and the same combined to make a floral design. It will be observed that the lines of this composition are strong and vigorous, corresponding with the character of Summer growth. In contrast to this, Fig. 25, on page 125, shows an arrangement of Autumn Chrysanthemums, which is altogether more mature in style, in accordance with the character of Autumn vegetation.

This plant requires great care in order to preserve freshness in the leaves. It should be cut after sunset, and the ends

Fig. 24.

of the cuttings should be burnt to charcoal, and placed for a considerable time in water before arranging, so as to prolong their vitality. The stems are brittle, and need great skill in bending to the desired curves, very refractory portions being softened by heat. The blossoms of the small Chrysanthemum, as shown in Figs. 24 and 25, are massed in groups, generally consisting of an uneven number of flowers. But with the species of large flower, the blossoms are isolated and of limited number, each stem bearing one, two, or at most three flowers. There are said to be seven faults which must be guarded against in the disposition of such large flowers. A blossom must not be turned away so as to present its back in the composition; nor must it turn its full disc to view; flowers must not have stems of the same length so as to be exactly in a line; nor must three be arranged in a triangular form; the disposal of blossoms in regular steps is objected to; *colour-sandwiching*,—as previously described on page 53,—is another fault to be guarded against; a large open blossom should not be put at the lower part of the composition; and flowers should not be hidden by leaves.

FIG. 25.

In addition to the special form of arrangement suggested by the peculiarities of growth at different seasons, three general styles of composition are recognized for floral designs; namely, the *finished* style, the *intermediary* style, and the *rough* style. These distinctions correspond with, and are described by, the same native terms as those applied to the square, and grass or running Chinese ideographs, and refer to different degrees of elaboration or *sketchiness*. In Plate XXXVI. are shown Chrysanthemums arranged in a *rough* style in combination with other flowers. Plate LI. exhibits two compositions of Chrysanthemums in a more *finished* style, one representing these plants disposed in five lines, with seventeen blossoms, in a fancy bronze vase, and the other showing a triple design in a bamboo vase of three mouths, with fourteen blossoms in all. A *rough* arrangement of these flowers in a special kind of basket is illustrated in Fig. 7, page 59.

ARRANGEMENT OF *NUPHAR JAPONICUM.*

The *Nuphar japonicum*, on account of its beautiful oval leaves, is a favourite water-plant for floral designs. It is arranged in *Sand-basins* and *Tubs*, the *Horse-tub* being by preference selected, and the *Horse's-bit*, or some other fancy fastener of metal employed. The largest leaves of this plant hang horizontally on their stems, and are gracefully curled at their edges like those of the Lotus. The stems are generally arranged as long as possible to convey the impression of a plant which grows in shallow water. This rule becomes specially important when the Nuphar is combined in a single vessel with other water-plants which flourish in less shallow pools; such deep-water plants being purposely kept as short as possible and placed more centrally so as to suggest the idea of greater depth. Though the flowers of this plant are small in comparison with the leaves, their position in a design is much studied, and the leaves are arranged in reference to them. Seldom more than two or three are introduced into a single composition, and these, consisting of a bud and one or two half open blossoms, are grouped between the *Principal* and *Tertiary* leaves.

For the *Principal* a large and slightly curled leaf is employed, placed higher than the rest and extending to the left of the design; for the *Secondary* a somewhat smaller leaf leaning to the right; for the *Tertiary* a small rolled leaf,—technically called the *Horn-leaf*,—is added below the *Principal*, and below this is a young bent leaf called the *Water-touching-leaf*, which should be only about an inch above the water surface.

The slender stems of this plant when arranged as above show a deficiency in width, and it is usual in broad water-vessels to introduce a secondary clump at the side, removed from the main composition by the space of a few inches. This may consist of a single flower and one small leaf, kept very short as if just sprouting from the water.

Double-well-buckets are often chosen for arrangements of the *Nuphar japoni-*

cum. In Plate XLVIIIA. may be seen such a composition in the lower of a pair of suspended *Well-buckets*, the Clematis being introduced into the upper one, in a hanging style, with a long *Streamer*. Another illustration (Plate XLIXB.) shows the same water-plant disposed with seven leaves and two flowers, and in Plate XXXVIIIA. it is combined in a large *Sand-basin* with a high group of bamboos.

ARRANGEMENT OF *KERRIA JAPONICA.*

This modest plant, with its long sprays of yellow blossoms, grows in abundance near the banks of rivers, and is specially associated with the scenery of the Tama-gawa, near Tokio. The accompanying illustration, Fig. 26, represents a composition made with the Kerria in a broad *Sand-basin*, intended to suggest such river-scenery. The stems are held by long cylinders of bamboo basket-work made in imitation of the *Ja-kago*, or monster baskets holding boulders which are used to break the current of rapid streams. Stones are added to complete the idea of the natural view.

Fig. 26.

The *Kerria japonica* is also frequently arranged in suspended receptacles and *Boats*, the stems having a leaning character given them with the object of preserving the notion of flowers hanging over and reflected into a stream. The popular fancy that this plant bears neither seed nor fruit, prohibits its employment at wedding ceremonies.

Plate XXXI. illustrates a defective and a corrected arrangement of the Kerria in standing vases, and Plates XXV. and LII. show the same flower disposed in a *Horse-tub* with the *Horse's-bit* used as a fastener.

ARRANGEMENT OF NARCISSUS FLOWERS.

The Narcissus takes an important place in floral arrangements during the Winter and Spring months. Available at a time of year when other blossoming plants are scarce, sweet in scent, and of golden colour, it is considerably prized by the flower-artist. Varieties of red, white, blue, and double blossom are said to exist, but with the exception of the white, which is sometimes used on ceremonial occasions, these rare specimens are seldom seen in floral compositions. The blade-shaped leaves of this plant receive chief attention; they are removed from the sheath, and are separately pressed and rubbed on a board to take out the excessive twist which the heat of the sun gives them in their natural state, a single curl alone being permissible for leaves employed in compositions. The younger leaves are easily smoothed and straightened with the fingers, but the large ones are refractory and require much labour and patience. Thus altered they are attached together in pairs consisting of one long and one short leaf, with the longer blade in front, and these pairs of leaves are employed for the different lines of the arrangement, with the flowers placed between. Leaves which curl over are only admitted in hanging compositions. The lengths of the flower stems are made to vary according to the particular season, being shorter in Winter than in Spring arrangements. If the blossoms droop they are carefully straightened and held upright by small wooden spikes. A simple five-lined composition with Narcissus is shown in Plate LIIIB. a double arrangement of Willow and Narcissus is illustrated in Plate XLA. and in Plate XXXVIA. may be seen a triple composition in a high bamboo vase in which the Narcissus occupies the central position, with Plum branches above and Chrysanthemums below.

ARRANGEMENT OF CHERRY BLOSSOMS.

The Cherry blossom, being regarded as the king of indigenous flowers, has precedence of all others for floral arrangements. It is considered preferable to use branches of Cherry blossom alone, and they are but rarely combined with other flowers. Sometimes, however, combinations with Pine branches, with Rape blossoms (*Brassica chinensis*), and with Irises are to be seen, though they are hardly considered *de rigueur.*

It is said that, in making arrangements with Cherry blossoms, the scissors must not be used. Though this rule is not strictly followed, the branches should be trimmed and bent as little as possible, and the composition partake more of a natural than an artificial character. In accordance with the distaste for over-exuberance in floral designs, the Cherry of small and single blossom is selected in preference to the richer specimens of double blossom.

A composition made with this flower should have a somewhat sparse arrangement of buds for the central line, half-open blossoms at the sides, and a few in full bloom below. Several fallen petals should be placed in the water of the vase, and any blossoms which have dropped may be left remaining on the flower stand. From the above rules it will be observed that the prevailing idea is to represent the Cherry-tree as nearly as possible in its natural state. The scenery of picturesque spots famous for the wild Cherry-tree, supplies the motive of certain compositions. In Fig. 11, page 67, is shown a simple bunch of blossoms arranged in a small flower-basket placed on a miniature raft, intended to suggest the flower-laden boats which float down the rivers at Yoshino and Arashiyama during the Cherry season. In Plate XLV. at D, is illustrated a composition made with sprays of the weeping Cherry placed in a hooking flower-basket.

ARRANGEMENT OF PEACH BLOSSOMS.

The Peach blossom can hardly be said to receive attention in Japan in proportion to its great beauty and richness. The Peach trees in flower form an attractive feature of the Spring landscape, but their gathered branches, though often employed in floral designs, are somewhat too exuberant for the taste of the flower artist. The Peach is used chiefly in flower compositions during the Girls' Festival of the third month. The cuttings are then considerably thinned out, most of the full-blown flowers being removed, and only a small number of half-open blossoms and buds left distributed in masses amongst the foliage. The double blossom flower is never employed, as it is said to have poisonous properties.

In Plate LIV. at A, is shown an arrangement of Peach blossoms in a standing flower-basket, together with the *Rosa indica*, and in the same illustration at D, the Peach is arranged alone in a hooking basket.

ARRANGEMENT OF CAMELLIAS.

The red and white Camellia, both single and double, are much used for Spring arrangements, though generally in combination with other trees or flowers. Their employment in conjunction with the Willow has been mentioned in discussing Willow compositions on page 116. There is a prejudice against the Camellia on account of a peculiarity of its blossom, which does not fall to pieces petal by petal like other flowers, but drops off bodily, suggestive, it is said, of a head struck off with the sword. On the other hand, it is held in considerable estimation as an indigenous tree which has been much used for various purposes from ancient times. It is also highly esteemed as an evergreen. In floral designs the disposition of the glossy oval leaves receives greatest attention; two leaves at least must be appended to each bud or blossom introduced into the design.

Sometimes these two leaves point upwards, extending above the flowers; sometimes they are bent over more horizontally, so that the whole of the blossom appears above them. The white Camellia, if in season, is much used at *Coming-of-age* celebrations. The favourite kind of composition is one of five blossoms and fifteen leaves, each flower or bud being surrounded by three curling leaves.

Examples of designs in which the Camellia is introduced may be seen in Plate XLIB. where it is combined with the Willow in a suspended bronze vase of crescent shape; in Plate XIVA. it is connected with the Peach and Narcissus. In Plate XLB. a particular kind called the *Camellia sasanqua* is shown, arranged with the Willow. Bamboo vases are specially suited for Camellias, those of the *Lion's-mouth* shape being selected by preference.

ARRANGEMENT OF CONVOLVULI.

The limp and delicate stems of creeping plants like the Convolvulus and Morning Glory are difficult to arrange according to the lineal rules of the Flower Art. Thin bamboo rods, withered stumps, or twigs are used as a basis in forming a composition, and the design consists in a judicious balancing of leaves, flowers, and buds, in which can be detected the radical elements of *Principal, Secondary*, and *Tertiary*. For the lower portion of an arrangement, flower buds are preferred to open blossoms. Rikiu is said to have originated a composition consisting simply of one flower and one leaf of the Convolvulus, which afterwards became a favourite decoration for *Tea-rooms*. Upon being asked why he adopted so modest a design, he replied that, as it was impossible to rival nature in her grouping, artificial arrangements should be as unassuming as possible; even a single flower with one leaf being sufficient, he maintained, to call for admiration. One is tempted to believe that the difficulty of arranging such unstable plants according to the accepted standards of lineal balance, may have had more to do with such a departure than the experienced master was ready to admit. The founder of the Enshiu School invented a method of employing the Convolvulus in a flower-basket, by winding the creeper round the long oval handle of the receptacle. Such a composition is illustrated in Fig. 6, page 58. A Convolvulus of three blossoms, placed in a standing vase on a high table, is shown in Plate LA.

ARRANGEMENT OF LESPEDEZA FLOWERS.

The Lespedeza is the principal of the Seven Plants of Autumn, and is much used for floral designs at this season. On account of its numerous small oval leaves and tiny flowers, a full and crowded arrangement of sprays is generally resorted to. When placed in standing vases a trilineal composition is followed, each *line* consisting of a group of four or five stems richly loaded with leaves and blossoms. The lower or *Tertiary* line may be composed of a number of short stems only sparsely supplied with leaves and flower buds, and curved in such a manner as to suggest the profile of a wild boar's head. The wild boar is supposed to sleep under the Lespedeza branches, and is associated in art with this Autumn plant. There is an early specimen called the Summer Lespedeza, and this should be arranged in a thinner and more open manner than the Autumn plant, in order to preserve its special character of growth, which is less profuse than that of the later season. In hooked or suspended receptacles, the Lespedeza is employed with a long *Streamer*, the other branches being kept short and disposed in a simple and quiet manner. Such an arrangement, placed in a crescent-shaped vase, is illustrated in Plate XLIIB. The Lespedeza is often arranged in suspended bronze boats. It is also sometimes used with the Morning Glory, the *Eularia japonica*, *Valeriana villosa*, *Valeriana officinalis*, *Pueraria thunbergiana*, and the Carnation, making altogether the combination called the Seven Plants of Autumn.

ARRANGEMENT OF THE *RHODEA JAPONICA.*

This water-plant is valued for its beautiful large leaves, which are arranged with the greatest care and precision, each important one receiving a special name according to its position or function in the design, as follows :—

The *Central-leaf,* which occupies the place of the *Principal* or middle point of the arrangement. The *Spring-leaf,* a young and curled leaf enveloping the central stem. The *Autumn-leaf,* placed in the background of the design and having a faded or withered tip. The *Dew-supporting-leaf,* employed in front of the *Principal* in Spring and Summer arrangements, and having a curl upwards. The *Dew-spilling-leaf,* used behind the *Principal* in Autumn and Winter designs, and having a curl downwards. The *Swallow's-mouth-leaves* consisting of two young leaves lapped together so as to present a double tip, and placed between the larger leaves of the composition. The *Frost-protecting-leaf,* which bends over some distance above the berries of the Rhodea, as if to shelter them from snow or frost. The *Wind-protecting-leaf,* which also curves over the berries, somewhat lower in position, as if to screen them from wind. The *Berry-protecting-leaves,* being two or three leaves disposed below and around the berries and from between which they appear to spring; in some arrangements these consist of withered leaves. It is considered imperative that, in floral designs made with this plant, the leaves should exhibit a proper balance of front and back surfaces. The colour and gloss, to which great attention is given, are much improved by sprinkling the leaves with *sake,* and immersing their stems for several hours in the same liquid.

In disposing the leaves in a composition they are piled in a step-like manner in pairs and threes. In this way, for a seven-leaved design, first a pair of leaves are placed, then another pair, and lastly three together; the berries are always intro-duced low down, screened and partly hidden by the foliage.

Arrangements of six different styles are recognized for the *Rhodea japonica.* as follows :—

The *Rainy-season-arrangement,* suited for any period of the year, in which style both the *Dew-spilling* and *Dew-supporting* leaves are employed. The *Snow-time arrangement* in which the *Frost-protecting* and *Berry-protecting* leaves are introduced, all withered leaves are discarded, and the berries are kept as low as possible in position. The *Spring-arrangement,* suited only for the early months of the year, in which many young leaves are used. The *Summer-arrangement*, distinguished by the introduction of a number of large leaves, with the addition of one withered leaf in front. The *Autumn-arrangement*, in which several withered leaves are placed and no young leaves allowed. The *Winter-arrangement*, consisting of a majority of withered leaves with the addition of the *Swallow's-mouth-leaves*.

ARRANGEMENT OF LEAF ORCHID.

The Chinese Orchid, called *Baran*, is one of the principal subjects for leaf arrangement in Japanese floral design, and its treatment serves as a model for most compositions with large-leaved plants. The flowers, being small and insignificant, are often omitted, but, when employed, they are attached by means of thin spikes of bamboo to raise them slightly in position. The *Baran* requires very careful treatment in order to preserve its freshness. It must be cut in the early morning or after sundown, and its leaves are then curled up, tied with string, and immersed in water for some hours before use. In very hot weather it is customary to suspend the cuttings for half a day in a deep well. To give a good colour and gloss to the leaves, *sake* is forced up their stems before immersing in water.

This plant is generally arranged in a water-basin, sand-basin, tub, or other wide-mouthed receptacle, with the addition of ornamental stones. In its natural state the Leaf-orchid has always one long oval leaf growing centrally and higher than the others, and in floral compositions this is used as the *Principal*, and is called the *End-leaf*. The bottom leaf of a series is small, with its point arching over; it corresponds to the *Tertiary* in floral arrangements, and is called the *Finishing-leaf*. Another special leaf introduced into certain compositions is one curled up spirally, and called the *Spider's-leaf*, being copied from leaves which are curled by the spinning of insects. This form is artificially produced by heating. A ragged leaf called the *Decayed-leaf*, made by tearing and scraping, is occasionally added. All the leaves of a composition including the above, are arranged in positions corresponding with the radical lines of a floral design, each leaf counting separately in the combination. Thus, in a three-leaved composition, the *End-leaf* will be used as *Principal*, the *Finishing-leaf*, *Decayed-leaf*, or *Spider's-leaf* as *Tertiary*, and an intermediary leaf will occupy the position of *Secondary*. In the same manner, for a larger composition of thirteen leaves, the *Finishing-leaf* is placed as *Principal* at the top and centre of the design; to the right and below are arranged four leaves called respectively *Secondary*, *Auxiliary* to *Secondary, Support,* and *Auxiliary to Support;* to the left are fixed four others, described as the *Tertiary, Support of Tertiary, Auxiliary of Tertiary,* and a *Decayed-leaf;* and along the centre, round the stem of the *Principal,* are placed four more, named in their order from above, *Support of Principal, Side-piece, Trunk-piece,* and *Auxiliary to Trunk-piece.*

All the stems of such a composition are closely united in a single line at the base for a distance of several inches above the surface of the water, and the leaves in most cases overlap one another considerably, only a few of the most important ones revealing as much as two thirds of their length. The distribution and balance of leaf surfaces receive considerable attention, so much so, that each leaf, in addition to its other names, is distinguished by the term *Front-surface-leaf* or *Back-surface-leaf.* All leaves are curled or twisted in some way, to show a portion of both sides, but rarely in equal degrees, so that a *Front-surface-leaf* would reveal only a point or edge of its back surface. In Plate XLIXA. this balance of surfaces is clearly indicated by shading. It is said that in a composition consisting of five leaves, three, including the *End-leaf* and *Finishing-leaf,* should be *Front-surface* leaves, but it appears that no strict rules are followed in this respect, a judicious balance and pleasing variety being alone sought.

Plate XXXIII. is instructive as showing defective and corrected arrangements of the Leaf-orchid side by side, in which not only the lines but the surfaces of the leaves are altered. An elaborate design, with the same plant in a hexagonal bronze vase, is illustrated in Plate XLIXA.

The Leaf-orchid is occasionally used in combination with other trees or plants, sometimes as the auxiliary and sometimes as the principal member of such double compositions. It is to be seen occupying a subsidiary position combined with the *Nandina domestica, Forsythia suspensa,* and large Chrysanthemum, and with the small Chrysanthemum and *Papaver rhœas* it holds the principal position.

ARRANGEMENT OF MAPLE BRANCHES.

The Maple, next to the Pine, is the most important flowerless tree used in Japanese compositions. Of this tree, there are two kinds,—the Spring Maple, which is red when the young leaves open, and the Autumn Maple, which is green in

Summer, and turns crimson later on in the year. Floral artists follow several fancy styles of arrangement with Maple branches, which are as follows :—

The *Sunrise-arrangement*, in which the leaves of the *Principal* branch should display their front surfaces. The *Sunset-arrangement*, in which the leaves of the *Principal* branch should have their under-sides turned to the spectator. The *Cloudy-weather-style*, in which the leaves should be curled and sprinkled with spray. The idea of this style of composition is taken from the appearance of the wild Maples of *Ogura-yama*, near Kioto, the leaves of which are often curled by frost. The *Tsuten-arrangement*, in which green leaves are used in the upper, and red leaves in the lower part of the composition. The name refers to a spot called Tsuten, famous for its Maple trees, whose leaves redden from below. The *Tatsuta-arrangement*, so called from a place of that name, where fine Maple trees line the banks of the stream. In disposing the *Principal* mass in this composition several of the larger leaves should be removed and placed floating in the water of the flower vase, to suggest the leaves which fall off into the river.

A divided arrangement of Maple branches in a large bronze bowl is illustrated in Plate X. also a combination with Chrysanthemums is shown in Plate XIVA. The Maple is often used in water-basins, and sometimes in combination with the Iris or other water-plants.

MISCELLANEOUS.

The above description of special arrangements with certain flowers includes those most often introduced into Japanese floral compositions. Nearly every tree and plant, however, the blossoms or foliage of which possess any beauty or attraction, may be seen occasionally introduced into designs, either singly or in combination. The manner of treatment and combination is based upon the principles already expounded as to characteristics of growth, locality, sex, and season, controlled in many cases by traditional fancies.

The examples of compositions with Leaf-orchids, *Rhodea japonica*, Lotuses, and *Nuphar japonicum* may be taken as models for arranging most plants having large oval leaves; the designs of Irises and Narcissus may be followed in employing plants having long, blade-like leaves; the arrangements of Clematis and Wistaria illustrate the manner of treating trees and plants of the creeper variety; the compositions with Plum, Cherry, Peach, and Willow branches serve as examples for the disposition of other straight-branched and blossom-clad trees; and the treatment of Chrysanthemums and Peonies can be taken as a guide for arranging most plants having large ponderous blossoms.

It only remains to allude to a few examples among the accompanying illustrations which have received no special notice in other parts of the work.

Plate LIVa. shows an example of the *Rosa indica* (Choshun) arranged as an auxiliary in combination with branches of Peach blossoms in a large flower basket.

Plate XXXVIb. illustrates the *Ilex sieboldi* (Ume-modoki), in a triple arrangement, placed in a high bamboo vase.

Plate XLIa. exhibits the *Tecoma grandiflora* (Nozen-kazura) in a suspended crescent-shaped vessel, balanced by a separate design of *Calendula officinalis* (Kinsen-kwa), in a standing vessel.

Plate XXXVIIb. shows the Cabbage-plant (Ha-botan) arranged in a globular standing vase.

Plate Lb. illustrates the *Dianthus superbus* (Nadeshiko) placed in a bronze vase.

Plate LIIa. represents the *Funkia ovata* (Giboshi), a large-leaved water-plant, in an arrangement of seven leaves.

Other plants occur in plates which are intended mainly to illustrate different receptacles for flowers. Among these may be mentioned:—the purple Magnolia, arranged in a bamboo *Roofed-boat*, Plate XXII. the *Patrinia scabiosæfolia*, in a wooden tub, called the *Long-boat*, Plate XXII. and again in a hanging vessel on a *Flower-horse* in Plate XXVI. the Vine, shown combined with small Chrysanthemums, in a crescent-shaped vessel attached to a *Flower-horse*, in Plate XXVI. Ivy with Camellias, in an inverted bronze umbrella, in Fig. 12, page 70; and the *Aster tataricus* paired with an arrangement of Irises in Plate XLVI.

PLATE XLVIII.

PLATE LI.

A

B

PLATE LII.

B

A

PLATE LIII.

A

B

<PLATE LIV.>

明治二十四年六月十日印刷并出版

所有
版權

發行所

兵庫縣播州龍野町　　　博聞社代理店

佐賀縣佐賀郡馬塲　　　博聞社代理店

福岡縣博多中島町　　　博聞分社

千葉縣千葉町　　　　　博聞分社

大坂備後町四丁目　　　博聞分社

東京京橋區銀座四丁目　博聞本社

印刷兼
發行者

兵庫縣士族

長尾景弼

一番地寄留
東京芝區西六保榮壽町

著作者

英國人

ジェー、コンダー

十三番地
東京京橋區西紺屋町

（定價金五圓）

www.ingramcontent.com/pod-product-compliance
Lightning Source LLC
Chambersburg PA
CBHW030643030726
47497CB00006B/1935